113140

INTERNATIONAL SERIES OF MONOGRAPHS IN
PURE AND APPLIED BIOLOGY

Division: **ZOOLOGY**

GENERAL EDITOR: G. A. KERKUT

VOLUME 51

PERSPECTIVES IN ZOOLOGY

OTHER TITLES IN THE ZOOLOGY DIVISION

General Editor: G. A. KERKUT

PERSPECTIVES
IN
ZOOLOGY

by

ALAN BOYDEN

Department of Zoology and
Bureau of Biological Research
Rutgers, the State University of
New Jersey

PERGAMON PRESS

Oxford · New York · Toronto
Sydney · Braunschweig

Pergamon Press Ltd., Headington Hill Hall, Oxford
Pergamon Press Inc., Maxwell House, Fairview Park, Elmsford,
New York 10523
Pergamon of Canada Ltd., 207 Queen's Quay West, Toronto 1
Pergamon Press (Aust.) Pty. Ltd., 19a Boundary Street,
Rushcutters Bay, N.S.W. 2011, Australia
Vieweg & Sohn GmbH, Burgplatz 1, Braunschweig

First edition 1973

Library of Congress Cataloging in Publication Data

Boyden, Alan, 1897–
    Perspectives in zoology.

    (International series of monographs in pure and
applied biology. Division: Zoology, v. 51)
    Bibliography: p.
    1. Evolution.   2. Zoology--Classification.
3. Phylogeny.   I.   Title.
QH366.2.B68 1973   591.3'8   73-1279
ISBN 0-08-017122-2

Printed in Great Britain by A. BROWN & SONS LTD., HULL

CONTENTS

PREFACE

To HAVE perspective is to have the capacity to see objects in their proper relation to one another, or the ability to judge the proper place and significance of facts and ideas. It is a capacity that will always be needed in biology, and it appears now to be lacking in the treatment of a variety of topics. In genetics, perspective is lacking where character expression is explained with reference only to genes, and to DNA and RNA, and no mention is made of the cytoplasmic organization. Perspective is lacking when evolution is viewed as concerned with the germ plasm only and not with somatic character expression. Perspective is lacking when evolution is defined in terms of change only, whereas in fact evolution is now concerned as much with the conservation of change as with the changes themselves. Recent studies have shown that many gene mutations have no effects on the proteins determined by the genetic code. This comes about because of what has been called the "degeneracy" of the code, in which several different triplet codons specify the same amino acid. Since the term "degenerate" implies that such codes have evolved from more perfect ones in which every different codon would specify a different amino acid, it is unwarranted in the absence of evidence that such evolutionary changes have occurred. In the meantime, we may refer to the code as ambiguous in such instances.

Perspective is lacking when it is assumed that the mechanisms of evolutionary change revealed in the study of highly specialized organisms are the only mechanisms which have ever existed, and that *known* simple genetic and reproductive processes played no part in primitive evolution. Perspective is lacking when it is believed that "the problem of systematics regarded as a branch of general biology is that of detecting evolution at work" (Huxley, 1940) and the great tasks of collecting, describing, naming, comparing and classifying the results of two or three billion years of evolution are slighted. Perspective is also lacking when it is claimed that classification should be related to "propinquity of descent" (Simpson, 1959) as though the mere passage

vii

of generations was the important thing in evolution and not the varying amounts and kinds of variation in the characteristics of the evolving organisms.

It is our plan to discuss in a critical way some of the aspects of biology where perspective is lacking and to call attention to the possibilities of obtaining a more correct view. It is not implied that all the points raised and suggestions offered will find immediate acceptance, or that they constitute the final word on the subject. Previous experience has shown that this is unlikely; but at the very least, attention will be drawn to the nature of the various problems and the need of further study and appraisal. Where the focus is sharpened, a clearer picture may emerge and the nature of the processes concerned may then be better understood. According to Singer (1959), history was originally an inquiry, and we shall open our discussion of natural history with some pointed inquiries.

It is a pleasure to recall many discussions of the topics touched upon in this little book, with colleagues and students at Rutgers University, at Kansas University and at Queen Mary College, University of London. In fact, these discussions took place during a period of several decades, and the views presented herein have been "naturally selected" with the aid of the searching and constructive criticisms of these stimulating and challenging critics. To all these my sincere thanks.

In this connection, special benefits have accrued to me from the receipt of two Senior Lecturer Fulbright Awards held at Queen Mary College in 1960–61 and 1966–67. During the first of these awards, a series of lectures entitled "Perspectives in Zoology" was presented to students and staff at Queen Mary College under the sponsorship of the Department of Zoology, Professor J. E. Smith, head. During the second award, a brief series of lectures in the field of Systematic Serology, shared with Dr. Mabel Boyden, was presented under the sponsorship of Dr. Gordon Newell, head of Zoology.

The appointment as Rose Morgan Visiting Professor to the University of Kansas during the Spring term of 1964 gave me another opportunity to discuss "Perspectives in Zoology" in the form of a Seminar, attended by graduate students and staff in Zoology. A small group of the very able graduate students of Dr. C. A. Leone provided searching criticisms to which satisfactory responses had to be made.

Finally a sudden recall to Kansas University in the Fall of 1965 to teach Genetics gave still another opportunity to refresh me in that field of study and benefit me with discussions among "The Evolutionists" and in Dr. Glen Wolfe's Genetics Seminar.

As anyone may plainly see, whatever of knowledge and understanding I may have acquired is polyphyletic in origin and in terms of the actual content of the following chapters most appropriately so. It is not possible to mention all those individuals and agencies from whom great benefits have been received during the long period (over ten years) of preparation for the writing of this book. But I must surely mention the Trustees of the British Museum (Natural History) in whose General and Zoological Libraries I have done much reading. Also, Dr. C. A. Wright, in whose laboratory of Experimental Taxonomy space was made available to carry on some serological testing and who prepared certain antisera for the purpose. Thanks are due also to the Trustees of The Linnean Society for the opportunity to use their library, especially to examine the writings of Alfred Russel Wallace, preserved in their possession. It is a pleasure also to thank Mr. and Mrs. A. J. R. Wallace for their gracious hospitality during my visit to "Wallace Country" (Bournemouth and vicinity), for the purpose of learning more about their illustrious grandfather. John and Daphne Wallace had arranged for Dr. and Mrs. Norman to attend a tea, at which Dr. Norman recalled his memories of Alfred Russel Wallace, whom he attended in his last years of life. I listened with the greatest interest to his recollections of the man and his works, and was given a tape recording by John Wallace so that Dr. Norman's recollections will be preserved. But we cannot go on to mention all who have aided us on the occasion of our visits to the Marine Biological Laboratories of the United Kingdom, or who have been our hosts during lecturing visits in Amsterdam, Brussels, Paris, Cork, and of course London.

Now we turn to more immediate persons, viz., to my daughter-in-law Nina Boyden whose expert typing has been of the greatest help. Without making any claims to understanding what I have written (Nina is completely non-biological) she has become expert at reading my writing, and is in fact the most rapid and accurate typist I have ever known. My sincere thanks to her for her help.

Last, but certainly not least, Dr. Mabel Boyden, my wife of nearly fifty years, has encouraged and helped me in all my undertakings. As Custodian of Collections in the Serological Museum, she has been indispensable, and as an active investigator in the study of lectins and blood groups, she has developed that part of the Serological Museum's programme. To her my words of thanks and appreciation seem totally inadequate.

ALAN BOYDEN,
"Redwood"
May 1, 1972

ACKNOWLEDGEMENTS

THE AUTHOR is indebted to the following holders of copyright for permission to reproduce published material.

Abelard-Schuman Limited: on p. 78, Fig. 4.1, from *A History of Biology* by Charles Singer, 3rd edition, 1959, Fig. 44, p. 91.

American Institute of Biological Sciences: on p. 132, Fig. 5.2, from "Homology and Analogy: a century after the definitions of 'Homologue and Analogue' of Richard Owen", *The Quarterly Review of Biology*, 1943, Vol. 18, No. 3, p. 238.

The American Midland Naturalist: on p. 129, Fig. 5.1, from *The American Midland Naturalist*, 1947, Vol. 47, No. 3, pp. 653 and 661, and quotations from "Some Meanings of Homology", p. 648, and from "Some Meanings of Analogy in Biology", p. 659.

The American Museum of Natural History: on p. 203, Fig. 6.12. No copyright.

N.V. Boekhandel en Drukkerij v/h E. J. Brill: on p. 63, Fig. 3.4, from *Acta Biotheoretica*, 1959, Vol. 13, p. 122.

Cambridge University Press: on pp. 95, 98, quotations from *The Evolution of Development* by J. T. Bonner, 1958.

The Clarendon Press, Oxford: on. p. 102, quotations from *The Organization of Cells* by Laurence Picken, 1960, pp. 92, 503–504.

Columbia University Press: on p. 71, quotation from *Trends in Genetic Analysis* by G. Pontecorvo, 1959, p. 1: on pp. 246, 253, quotations from *Principles of Animal Taxonomy* by G. G. Simpson, 1961, p. 91.

Marine Biological Laboratory: on pp. 170, 171, Figs. 6.9, 6.10, from the report by Boyden, "The precipitin reaction in the study of animal relationship", *The Biological Bulletin*, Vol. L, No. 2.

McGraw-Hill Book Company: on pp. 6, 7, 10, quotations from *The Invertebrates* by Hyman, Vol. I, 1940, pp. 5, 44, and Vol. V, 1959, p. 710. On pp. 234, 242, quotations from *Principles of Systematic Zoology* by Ernst Mayr, 1969, pp. 2, 85.

Rutgers University Press: Plate Fig. 6.8, from report by J. Munoz, p. 66, The use and limitations of serum–agar techniques in studies of proteins. In W. H. Cole, editor, *Serological Approaches to Studies of Protein Structure and Metabolism*, 1954.

Scientific American: on p. 37, quotations from "The Oldest Fossils" by Elso S. Barghoorn, 1971, pp. 30 and 41. Copyright © 1972 by Scientific American, Inc. All rights reserved.

Charles C. Thomas, Publisher: on pp. 31, 33, Figs. 2.1, 2.2, from *Protozoology* by R. R. Kudo, 4th edition, 1960, Figs. 70, 71, pp. 172, 173.

University of California Press: on pp. 218, 219, quotations from *The Triumph of the Darwinian Method* by Michael T. Ghiselin, 1969.

The University of Chicago Press: on pp. 60, 62, Figs. 3.2, 3.3, quotations from *Animals without Backbones* by Ralph Buchsbaum, 1938. On p. 147, Figs. 6.2, 6.11, from "Precipitins and Phylogeny in Animals", *American Naturalist*, 1934, Vol. 68, Figs. 1, 2. On p. 137, quotation from "Evolution and Morphologically Uniform Groups" by W. J. Bock, *American Naturalist*, 1963, Vol. 97.

University of Utah Press: on p. 87, quotations from chapter by J. B. Gurdon, p. 225. In *Problems in Biology: RNA in development*, edited by E. W. Hanley.

University of Wisconsin Press: on pp. 96, 97, quotation and Figs. from chapter by J. B. Gurdon, p. 220, and chapter by T. M. Sonneborn, Figs. 19.2 and 19.3, pp. 388 and 389, *Heritage from Mendel*, edited by R. G. Brink (Madison: The University of Wisconsin Press; © 1967 by the Regents of the University of Wisconsin).

Yale University Press: on pp. 19, 20, quotations from *The Planets* by H. C. Urey, 1952.

CHAPTER 1

NATURAL AND UN-NATURAL HISTORY

Is natural history now really natural?

THE standard histories of biology, such as Nordenskiold (1928), Singer (1959), and Dawes (1952), give the names and dates and many particulars in regard to the chief persons and events which have been concerned in the development of the science of biology. From them we can get some appreciation of the slow movement, and sometimes meandering course, of the stream of knowledge and understanding. One can also see that there has been a trend away from a subjective and supernatural interpretation of the structures and functions of living things. But has this progress gone far enough? Is natural history now as objective and natural as it should be?

In my opinion, Darwin's greater contribution to biology was that he made it possible for evolution to be understood as a natural process, rather than a supernatural one. This accomplishment is not lessened by any of the criticisms of Darwin, mentioned by Darlington (1959), or by Eiseley (1959), however valid they may be. In fact, the controversy which developed after the publication of *The Origin of Species* was concerned more with this aspect of evolution than with the validity of natural selection *versus* alternative explanations for its occurrence. This was all to the good in the sense that greater understanding of life resulted from the more natural view of evolution, but there was, and still is, an unfortunate consequence of Darwin's work in that the direction of biological thought and enquiry was directed backwards toward ancestry rather than toward the nature of the organisms themselves.* In fact, many biologists appear to belong to a kind of cult of ancestor worshippers* subscribing to a host of unproven and oftentimes conflicting views in regard to the remote ancestry of groups of organisms.* This general tendency for the acceptance of assumptions

* See Kerkut (1960), *Implications of Evolution*, chapter 1 and following, for a critical account of these matters.

regarding origins as proven facts has seemed to carry with it an attitude of carelessness in the capacity to distinguish fact from theory, and in the ability to classify biological structures, functions and processes in accordance with their natures, instead of on the basis of assumed remote ancestry.* In these respects, modern biology has become less natural than it should be, and less scientific. Biologists have need to be at least as critical and objective as any other scientists, and their thinking must be no less rational and logical.† Let us first consider some instances where logic and reason appear to have been set aside in the treatment of particular problems.

The real nature of parthenogenesis

Is parthenogenesis sexual or asexual reproduction? Some years ago, I asked this simple question (Boyden, 1950) and attempted to answer it in a reasonable way in accordance with the natures of the two kinds of processes. So far as I know, there have been no other published answers, and it is still correct to say that some principal texts of cytology, Wilson (1925 and later editions), Darlington (1958) and White (1954), treat parthenogenesis as "asexual reproduction" and thus group it with the processes of budding and fission, to which it has no real similarity or relationship. If sexual reproduction is defined as "reproduction which involves gamete formation" (Boyden, 1950) the process is characterized in terms of a constant criterion, and not in terms of some highly varied and uncertain assumed consequence which may result from fertilization. For the consequences of fertilization depend on the breeding system and there may or may not be biparental

---

* It is interesting that Darwin's monographs on the subclass Cirripedia, published in 1851 and 1854, were considered by him to be "strictly systematic" works. Though his barnacle work was begun after he had written his two sketches of his evolutionary theory he classified all the species available to him on the basis of their characteristics as he observed them, and made no claims in regard to their genealogic relationships. By using the morphological and life history characters revealed to him during eight years of intense and exhausting study, he was able to develop a classification which has stood up for over a century. (A more complete discussion of these matters will be found in our Chapter 7.)

† My use of the word logical is a broad one, for it assumes that there is an implication that the premises selected are in accordance with existing knowledge. I do not consider it "logical" to adopt false and arbitrary premises even though you stick by them to the bitter and useless end.

inheritance with recombination and increased variability, or heterosis, etc.

Those who say that parthenogenesis is "asexual" base their decision *on what happens to an egg after its formation* and ignore the long series of ontogenetic and phylogenetic events that have led to the production of such highly specialized cells. No other cells are prepared to do so much as are egg cells for no other cells are regularly called upon, in metazoa, to produce whole organisms. And no other cells but the gametes are produced in such animals by those precise and complicated steps which occur in gametogenesis. To classify parthenogenesis as "asexual" ignores the true nature of the process, and misrepresents its evolutionary history. All cytologists seem to be agreed that parthenogenesis is a derivative of normal sexual reproduction and that such a derivation has probably occurred separately in a number of distinct animal groups.

The hive bee (*Apis mellifica*) may serve as an example of one type of parthogenesis. Here all eggs are haploid and produced in the typical way. If the queen has been mated successfully, sperms will be available and whether they gain access to the eggs and fertilize them depends upon the operation of the female reproductive organs. The eggs which are fertilized are thus made diploid and become queens or workers. On the other hand, the unfertilized eggs develop into males or drones, which remain haploid, and which will eventually produce sperm by a suitable modification of the usual gametogenetic processes.

To sum up, eggs, produced in ovaries and housed in the female reproductive tract, may be fertilized by sperm produced in testes and housed first in the male and later in the female reproductive tracts. If fertilization occurs females are produced, and if the eggs develop without fertilization, males develop. The *whole* cycle of events is characteristic of the nature of sexual processes and there is no trace of anything comparable to simple budding or fission. To refer to the parthenogenetic development in this cycle or any other as asexual, which properly means without sex, can only misrepresent both the nature and the evolutionary derivation of the process. This is a good illustration of un-natural even sub-natural history!

Now it may be objected that although in such cases of haploid parthenogenesis as occur in bees (*arrhenotoky*) there is no trace of asexual processes, the situation is different where diploid or polyploid parthenogenesis occurs. Here the eggs might be considered as ordinary somatic cells and their division as a kind of fission process. But this is not really the case in any kind of parthenogenesis in which females are produced (*thelyotoky*). White (1954) divides such cases into two main kinds, (1) the meiotic type in which meiosis results in a reduction of the chromosome number, which is returned again to the somatic number by a doubling of the chromosomes at some stage in the life cycle, and (2) the ameiotic type in which meiosis has been entirely suppressed, the maturation division or divisions being mitotic in character. But even in the ameiotic type of thelyotoky, typical gameto-genetic processes occur up to the maturation period of gametogenesis and the cells involved therefore are in no case ordinary somatic cells.

These modifications of the machinery of gamete formation have apparently evolved independently from processes of normal sexual reproduction with fertilization, in many specialized groups of Metazoa, but never by any kind of modification of budding or fission processes. They reach various degrees of permanence in different species. Thus in the plant lice parthenogenesis occurs in spring and summer and return to the production of haploid gametes, fertilization and zygote or "winter egg" production occurs in the fall. In aphids, rotifers, phylloxerans, daphnids and ostracodes, parthenogenesis occurs as the major form of sexual reproduction, with only occasional return to generations resulting from fertilization. A further evolution in the direction of uniparental inheritance occurs when males are extremely rare or entirely unknown as is the case in some Nematoda, some species of Phasmidae, Tenthredinidae, Coccidae and Physidae.

According to Wilson (1925) the evidence makes it highly probable that in all these cases the original mode of reproduction was normal sexual and that it has been supplanted by parthenogenesis. This change seems to have taken place readily, for even within the limits of a single genus some species may reproduce only by normal sexual process, others by parthenogenesis only, still others by both. Again, after reviewing the information available from a considerable number of studies concerned with the details of natural and experimental partheno-genesis, Wilson (*op. cit.*) concluded that to the cytologist the processes called forth by fertilization or parthenogenetic activation offer the appearance of a train of connected events, more or less plastic in each individual case and varying materially in its details from species to species.

Now in spite of all the evidences discussed by Wilson regarding the derivation of parthenogenesis from normal sexual reproduction with fertilization he persisted in referring to diploid parthenogenesis as "asexual". That this characterization is inappropriate and misleading is shown by the following facts:

1. Haploid parthenogenesis is concerned with typical gametogenetic (sexual) processes throughout.

2. Diploid parthenogenesis results from typical gametogenetic processes in many species up through the formation of the first polar body, the second meiotic division being suppressed and the secondary oocytes thus remaining diploid. However the formation of even a single polar body is enough to show that this is a gametogenetic process. In other cases the maturation division or divisions may be mitotic, but the stages prior to the maturation period have been typical for gametes, not ordinary somatic cells.

3. These phenomena of parthenogenetic development occur most commonly in specialized groups of Metazoa, in no one of which has typical asexual reproduction (budding or fission) ever been observed.

4. In Metazoa, budding and fission are usually concerned with multicellular processes or regions of the body and not with the production of single reproductive cells whether haploid, diploid or polyploid.

Neither in nature nor in evolutionary derivation, therefore does parthenogenesis show any relationship to the known types of asexual reproduction and it is arbitrary, un-natural, and misleading to classify it in any such way. The truth of the matter is that there are no apparent transitions between sexual and asexual reproduction anywhere in Metazoa, the processes being distinct throughout.

Are there any "acellular" animals?

I have been concerned about some unfortunate references to Protozoa as "acellular" animals (Boyden, 1957a, b). Dobell (1911) in no uncertain terms claimed that Protozoa must not be considered to be cellular organisms. Hyman (1940) has taken a similar stand in

Volume 1 of her monumental series, *The Invertebrates.* To a zoologist who believes with Owen (1866) that "terms are the tools of the teacher, and only an inferior hand persists in toiling with a clumsy instrument when a better one lies within his reach" this usage has been disturbing.

Both Dobell and Hyman agree that individual Protozoa have the same funamental organization as is to be found in the cells of Metazoa. In spite of this Dobell states definitely that the protist individual is not the homologue of a single cell in the body of a multicellular animal or plant, but is instead "homologous" with a whole multicellular organism. This claim discloses a fundamental misconception of the nature of homologous correspondences. Homology, whether serial or special, is a correspondence among *parts of organisms* and not, properly speaking, between whole organisms. It is incorrect then, as well as misleading, to attempt to homologize a protist with a whole Metazoan, for the relation of homology can only exist between the parts of a Protist and the corresponding parts of a Metazoan. Those corresponding parts are, obviously, the essentially similar parts of the cell or cells of each type of organism. Nuclei are the homologues of nuclei and cytosomes are the homologues of cytosomes, and many lesser structures present in both Protozoa and in the cells of Metazoa are also homologous. If these homologies exist, it is wholly improper and incorrect to refer to Protozoa as acellular and Metazoa as cellular, which usage would imply a fundamental difference in the protoplasmic organization of the two types of organism.

Not only has Dobell erred in the way in which homology was understood, but both Dobell and Hyman have been unfortunate in their arbitrary definitions of a cell, express or implied. Thus, Hyman* (1940, p. 5) says, "A cell is commonly defined as a mass of protoplasm containing a nucleus but would be better described as *one nucleated division of an organism.*" The intention of this definition was apparently to insist that no organism could be cellular unless it was *multi*cellular; and that a cell could never be a whole organism but only a part of an organism.

This, it seems to me, is a wholly arbitrary restriction on the meaning of the term "cell", which misrepresents the nature of the essentially

---

* By permission from *The Invertebrates*, Vol. 1, by Hyman, L. H. Copyright (1940) McGraw-Hill Book Company, Inc.

similar protoplasmic organizations found in both Protozoa and Metazoa. The use of such a definition leads to serious difficulties as is evident in the following examples. According to both Dobell and Hyman, a fertilized egg is not a cell, not because it has any non-cellular organization, but because it represents a whole organism at that early stage of development, and they insist that a whole organism must have many cells—or none! It does not matter to them that after the first cleavage the organism suddenly becomes cellular without any essential change in this protoplasmic organization! Gametes are called cells because they are interpreted as parts of organisms; hence an egg that develops parthenogenetically begins a life-cycle that is cellular throughout, but if the egg has been fertilized, it would immediately become non-cellular, to be followed by cellularity in the two-celled and subsequent stages! There could be no better illustration of the fantastic consequences of an arbitrary definition, unrelated to realities in nature. There is simply no justification for imposing on nature the arbitrary dictum that a single cell cannot be a whole organism, which is contrary to fact. What was hoped for from such unfortunate concepts of Protozoa and cells? Dobell wished to emphasize that an individual Protist could act as a whole organism, just as a Metazoan could. He also wished to make it clear that the Protista are not necessarily simpler, lower, or more primitive than Metazoa. With all this, I am fully sympathetic, but I will not agree that it was necessary to consider the Protozoa as "acellular" in order to accomplish this purpose, or that is is proper to speak of "homologizing" a Protozoan with a whole Metazoan organism. Admittedly whole organisms, Protozoan and Metazoan, may indeed be *comparable* and even essentially similar in respect to many aspects of their individuality and their living.

Hyman (1940, p. 44) gives a somewhat similar explanation of her position.

"The Protozoa are not loose cells moving about, but complete organisms that may be of more complicated construction than the simplest Metazoa. We therefore prefer to refer to the Protozoa as *acellular* rather than as unicellular, animals, that is, as animals whose body substance is not partitioned into cells."

I answered the question "Are there any acellular animals?" (Boyden, 1957a) in the negative for it was maintained that in view of the essential homologies between Protists and Metazoan cells, admitted by all those who claimed Protozoa to be "acellular", that we have no real alternative but to accept the cellularity of the Protozoa. This conclusion was also that of Minchin (1916) who criticized Dobell for failure to act in accordance with the homologies concerned. Minchin properly considered that Protozoa and Metazoa could be considered to function as individuals in an *analogous* manner—"a comparison, however, which leaves the question of genetic homology quite untouched". I certainly agree in principle with Minchin that "the view generally held that the entire organism of a Protozoan is truly homologous with a single body cell of a Metazoan seems to me quite unassailable". However, it would seem more correct to reserve the term homologous for the parts of the cells in each case.

The question of cellularity in Protozoa was discussed by others, following my report. Corliss (1957) stated that probably most proto-zoologists would agree with me in the conclusion that Protozoa are cellular organisms, but he considered that my "recent outburst of criticism directed at Dobell and Hyman . . . seems rather harsh and narrow", and that my account was deficient in that no precise definition of a cell was given.* The latter point is certainly correct, for I was mindful of the difficulties of defining a cell where, even though the central theme is consistent and recognizable, the variations in expression cover a considerable range. One could define a man, for example, in such a way that a one-legged man would become non-human! This I would avoid at all costs. Corliss goes on to quote J. R. Baker's (1948) definition of a cell, which, however, neither he nor I could accept as useful. According to Baker a cell is "a mass of protoplasm, largely or completely bounded by a membrane, and containing within it a single nucleus formed by the *telophase transformation* of a haploid or diploid set of anaphase chromosomes". But this definition also has an unfortunate and arbitrary restriction in that if there is one nucleus with a cytosome you have a cell, but if you have more than one nucleus (a purely quantitative difference) you have *no* cell! Surely it is not justifiable to create a qualitative and far-reaching distinction between protoplasmic organizations of the same fundamental kind, differing only in number of parts or in degree! If you must

* Neither did Corliss provide a definition of a cell!

define a cell in terms of each possessing a single nucleus, then the ciliates would often be bicellular, or in some cases multicellular, but *not acellular*!

I differ with Corliss in his narrow conception of what is "logical" for he states that Dobell's conclusion that Protozoa are non-cellular "represents a stand not at all inconsistent or illogical, it seems to me, with respect to his own definition of a cell". And again, bearing in mind Baker's definition, Corliss continues, "Thus it need not be considered altogether illogical to think of the protozoa as comprising a variety of forms some of which are clearly only unicellular, others multicellular in certain stages, still others acellular throughout their lives. Personally, however, I favour rejection of the circumscribed definition of a cell offered by both Baker and Dobell, and I consider the protozoa, as a group, to be unicellular organisms (not necessarily animals)." But where is the logic, I may ask, of pointing out that Baker and Dobell (and Hyman) were logical in their conclusions, and then rejecting these conclusions? Is it really "logical" to build up a case on false premises? My own interpretation of logic is a broader one and there is a require-ment of rationality built into it. It is not logical, in my view, to erect arbitrary definitions of such a kind as to lead to a denial of real homologies where they exist, or to create an illusion of a qualitative difference in protoplasmic organization where it is a matter of degree only (e.g., one nucleus or two).

The comments of Hutner and Provasoli (1957) reveal a failure to understand the nature of homologous correspondences. They say:

"We consider that protozoa are homologous in their general struc-ture to the cells of metazoa but that they have the autonomy of the whole organism. We contend, therefore, that it is necessary to be clear about where the homology of cell to cell ends and where the homology of cell to organism starts."

There is no such thing as the homology of cell to organism, except in terms of the corresponding parts of each. It is unfortunate that the whole discussion has been confused in this way, by bringing in the "concept of organism" as though it were opposed to that of homology or cellularity. I have never denied that Protozoa were organisms, nor that they were important and sometimes exceedingly complex. It is a very strange thing, and a witness to the confusion in thought involved in this discussion, when Hutner and Provasoli report, "Boyden states that 'because of their essential (*sic*) correspondence part for part with the cells of Metazoa, Protozoa are undoubtedly cellular', and thereby begs the question." On the contrary, the question was a relatively simple one and it seems that I was the only one who really stuck to it and did not "beg it".

Hyman (1959) also refers to the "concept of organism" in the Retrospect of Volume V. "The concept of the Protozoa as acellular organisms met with the usual resistance of inertia. The extreme of a narrow rigid view is expressed by Boyden (1957a), and in fact this author appears unable to grasp the concept of organism. The temperate, well considered reply of Corliss (1957) admits that some Protozoa conform to the definition of a cell and others do not." What Hyman fails to point out is that Corliss said that *on Baker's definition of a cell* some Protozoa would be considered cellular and some noncellular, but that he (Corliss) rejected such circumscribed definitions as those of Dobell and of Baker and considered the Protozoa as a group to be unicellular organisms. Miss Hyman has thus succeeded in giving an erroneous impression of Corliss's view, which was then, and still is, definitely opposed to hers (Corliss, 1959). She continues by stating that "the claim of Boyden that there is 'essential structural correspondence part for part' *is simply untrue*" and offers in support of this statement a list of organelles in Protozoa which have no counterpart in Metazoan cells. Now it must be obvious to most biologists that an organ or part in one organism cannot be homologous with *no* organ or part in another, and I spoke of the homology between the corresponding parts "present" in each. The lack of such parts has no direct bearing on the homologies which do exist between the common parts. Finally, she concludes "obviously there is only one way of regarding Protozoa that will cover all facts about them, that is, as equivalent, not to a metazoan cell, but to an entire metazoan organism. The terms acellular and noncellular are probably unfortunate, and it is hoped better ones may be suggested."

The quotation just above introduces a new term "equivalent", which has definitely not the same meaning as homologous and thus further complicates the discussion. But I will suggest only that I agree with Miss Hyman that the terms acellular and noncellular were unfortunate in view of the fact that all the while the terms cellular and unicellular were lying close at hand.

Now the comment of Martin (1957) may lead to a positive bit of progress in this controversy. Martin calls attention to the organization of plasmodia. In some cases as the myxomycetes, there may be large numbers of nuclei embedded in a common matrix. He points out that it is difficult to consider this as a single cell, or even as a multicellular state and it may therefore be better termed acellular. My own conclusion was that the plasmodia could be termed cellular, inasmuch as nuclei and cytosomes still exist in a mutually interdependent relationship such as is always characteristic of cells; and many gradations between a uninucleate and extremely multinucleate condition may be seen. It is, to me, a question of difference in quantity again, rather than a qualitative difference, though perhaps Dr. Martin would consider that the apparent lack of any *constant* relation between each nucleus and a given portion of cytoplasm invalidates the use of the term cell.

In a lecture dealing with these subjects at Queen Mary College, University of London, I was asked by a student for a definition of cell and not permitted to excuse myself from giving it. After some reconsideration it has seemed to me that the definition I gave avoided the arbitrary restrictions which have been previously referred to, and yet described the kind of protoplasmic organization which most biologists would recognize as cellular.

"A cell is a unit of that kind of protoplasmic organization in which there is a mutually interdependent relationship between an internal phase of nucleoplasm, often consisting of one nucleus, and a continuous external phase of cytoplasm."

This definition would, I believe, cover the kinds of protoplasmic organization which are of one major kind, i.e. "cellular", and not be subject to unwarranted restrictions. Above all, it does not lead to the absurd conclusions that if an organism cannot be divided into cells it is not cellular (Hyman, 1940). As a matter of fact, no cell can be "further divided into cells", and if this were the decisive criterion for cellularity *no cell would be cellular*!

Now it would be expected that further knowledge gained from the study of the fine structure of organisms by the procedures of electron microscopy would bear on this matter. It certainly does! Pitelka (1963) has a chapter entitled "Protozoa as Cells" in her recent book, *Electron-microscopic Structure of Protozoa.* Two sentences will make clear her conclusions. "The electron-microscope has demonstrated that the fine structure of protozoa is directly and inescapably comparable with that of cells of multicellular organisms." And, "The morphologist has to start out by admitting that protozoa are, at the least, cells" (p. 8).*

Finally, reference should be made to the discussion by Gregg (1959), who approaches the problem of "deciding whether Protistans are cells" from the logical and philosophical points of view. He refers to "the Wilsonian view, namely the view that all organisms satisfy either the term 'cell' or the term 'cellular' ". His conclusions are, briefly stated, that the Wilsonian view is consistent whereas some of the opposing views are logically or empirically indefensible. Also the Wilsonian view is no less precise than those that have been put forward to replace it, and finally, "laws may be formulated in Wilsonian terms that are matched in generality, simplicity and utility by none belonging to any alternatives to Wilson's view that have been thus far advanced".

It is in some respects unfortunate that this whole discussion became so complicated by the addition of irrelevant matters, but it does provide a further instance of what I would consider "un-natural history". My hope is that biologists will be as critical and objective as any other scientists and that they will be expected to use both logic and reason in their definitions and expositions. It should not be left to the editors, or to the referees, to point out to the expectant authors that whole sections of their writings have no relevance to the topics being discussed, but there should be a body of opinion in the biological world that would definitely reject such "beating around the bush". In the words of both good teachers and good students, "Answer the question if you please."

These are by no means the only parts of zoology which may be considered un-natural. The fantastic redefinition of homology in terms of ancestry alone, leaving no terms whatever with which to refer to essential structural correspondences, is another such case (Bock, 1963; Mayr, 1969). As is also the claim that the appraisal of such similarities can only be subjective or intuitive whereas the criterion of common

---

* From Pitelka, D. R. *Electron-microscopic Structure of Protozoa*, 1963, copyright Pergamon Press, New York and London. With permission.

ancestry is theoretically objective! (Simpson, 1961). Or that the "pure phylogenetic system" is the only system "where the speculative element is absent or at least infinitely small . . ." (Kiriakoff, 1962).

Another case is supplied by those who talk about the "emergence of asexuality in the Protozoa" (Hawes, 1963) or claim that sexual reproduction was the original and universal process to be lost in some lines of subsequent evolution (Dougherty, 1955) when both the distribution of asexual reproduction and the nature of the process are such as to indicate that they are the more primitive genetic mechanisms in Protozoa and Lower Metazoa. But the treatment of these topics will be given in the places where the situations may be more adequately presented, that is, in the chapters which follow.

CHAPTER 2

THE GREAT AGES OF EVOLUTION

Has there been an evolution of evolutionary mechanisms?

PERSPECTIVE has been lacking in the study of animal evolution. Current views in regard to the nature of evolutionary mechanisms are based largely on the genetic behaviour of highly specialized and sexualized organisms, themselves the results of hundreds of millions of years of evolutionary variation, selection and fixation. It is to the birds and the bees and to mice and men, and of course to Drosophila, that geneticists have turned for information in regard to "evolution as a process". Even when they turn to "lower organisms", such as Bacteria and Protozoa, it is the more specialized and complicated genetic and reproductive processes in these species, to which they give their attention, rather than to the simpler asexual reproductive processes which they still retain.

In consequence, it is apparently the current view that all evolution is dependent upon shifting gene frequencies in cross-fertilizing organisms, and that it is safe on this assumption to extrapolate back from the tips of the evolutionary trees to the very roots. Obviously, this approach is based on the assumption that there has been no evolution of evolutionary mechanisms in two billions of years of evolutionary experimentation, but only in the products of these mechanisms. This is a most unlikely assumption.

It is further evidence of this lack of perspective when so many current discussions of phylogeny are based on the assumption that the *original* steps in the evolution of the major types were made by means of *modern* specialized genetic machinery, and no thought is given to the possibility that the simpler reproductive processes *now known* might have played a part in these early developments. And even in some general texts of zoology, the treatment is such as to indicate that asexual reproduction is considered to be of no evolutionary or general biological significance. This attitude has had a long, if not venerable, history. Harvey is acknowledged as the author of the aphorism "*omne animal ex ovo*" (Willis, 1847). According to Meyer (1936) the correct quotation is "*Ex ovo omnia*". In either case the general inference is the same, viz., that all reproduction is sexual and no other kind is worthy of discussion. As we shall see later this approach has led to the ignoring of facts which may have a very important bearing on primitive evolution.

It is our view (Boyden, 1953 a, 1954; Boyden and Shelswell, 1959) that there indeed has been an evolution of evolutionary mechanisms, as well as of the products of such mechanisms, and that during the great ages of animal evolution these mechanisms have evolved from primitive to more advanced states, from simple and generalized to specialized conditions, and that this evolution has occurred at times in divergent, or parallel, or even convergent pathways.

Any complete theory of animal evolution must concern itself with *all* the kinds of mechanisms concerned and all the periods during which they operated, viz.,

1. The period in which the conditions were set up so that life could originate.

2. The period in which there appeared the first living things, and whether they were heterotrophic or autotrophic; monophyletic or polyphyletic; and especially their primitive genetic and reproductive machinery.

3. The period in which evolutionary processes developed which culminated in the perfection of the reproductive mechanisms so that mitosis, meiosis and sexual specializations were achieved.

4. The period in which both the nature of the organisms and their genetic mechanisms were stabilized. This period includes at least the last 600,000,000 years.

The beginnings

Considerable attention has been given in recent years to the problems associated with the origin of life on earth. The recent revision (1957) of

Oparin's *The Origin of Life on the Earth* provides but one of several sources which present critical summaries of present facts and theories. There is also the extensive report (1959) of the Russian Conference on *The Origin of Life on Earth* edited by A. I. Oparin and others. Florkin (1960) has edited a selection of the reports from that Russian Conference under the title *Aspects of the Origin of Life*. Nigrelli (1957) has edited the report of the New York Academy of Sciences "Modern Ideas on Spontaneous Generation". Hans Gaffron contributed a chapter on the "Origin of Life" to Volume I of *Evolution after Darwin* edited by Sol Tax (1960). Further reviews and critical accounts have been written by Gösta Ehrensvärd (English translation, 1962) and John Keosian (1964). These together with some of the more significant original papers are the sources upon which our own discussion is based.

Perhaps the more suitable way to proceed will be to ask some specific questions and then to present the facts and interpretations which appear to make it possible to give some tentative answers. Among these questions are the following:

1. How old is the earth?

2. How was it formed, and what of the physical and chemical conditions of the earth's early surface? What of the temperature, amount of water and composition of the atmosphere?

3. What were the steps in the natural chemical evolution of substances necessary for the synthesis of protoplasm? When may life have first appeared and was it the result of a "rare accident" or a part of the processes of natural chemical evolution which occurred wherever and whenever the necessary conditions existed?

4. What was the probable nature of the first living matter and was its nutrition heterotrophic or autotrophic and anaerobic or aerobic?

5. If, as appears probable, the earth's primitive atmosphere was reducing, when and by what means did the change to an oxidizing atmosphere occur?

We shall proceed now to discuss some tentative answers to the above questions, as well as the implications of these answers for those who wish to gain some further appreciation of "Perspectives in Zoology".

The age of the earth

Current estimates are fairly consistent in their order of magnitude. Urey (1952), assuming that the age of meteorites may represent that of the solar system, reported that many, the ages of which were estimated by a variety of methods, were older than 3×10^9 years. Later (1963) he reported the age of the Canyon Diablo meteorites to be about 4·5 to 4·6 $\times 10^9$ years, believing that these estimates were the first which seemed to be reliable.

For the earth itself, Ehrensvärd (1962) and Vinogradov (in Florkin, 1960) give the age as 5×10^9 years. Urey (1963) gave the age as 4·5 $\times 10^9$ years. More recent rock-dating gives the figure of 4·0 $\times 10^9$ years for the oldest rocks of western Greenland.

For the moon rocks, the estimates are of the same order of magnitude and quite consistent. Silver (1970) reports that four lunar rocks have $^{207}Pb/^{206}/Pv$ ages in range of 4·13 to 4·22 $\times 10^9$ years, and that these are at least 400 million years younger than the dust and breccia samples. The latter have ages of 4·63 and 4·60 $\times 10^9$ years. It is stated also that these rocks are older than any known terrestrial rocks, but that cannot be by very much if we bear in mind the estimated age of 4·0 $\times 10^9$ years for the west Greenland rocks reported in a television presentation sponsored in the United States by the Ciba–Geigy Corporation and entitled "The Restless Earth".*

How was the earth formed?

Theories in regard to the manner of formation of the solar system and the earth have undergone significant changes in recent decades, and clear statements in regard to a recent theory may be found in Urey (1952a, 1963). In brief, this theory assumes that the solar system was produced from a cosmic cloud of cold dust and gases, which contracted into several large bodies, including our sun and planets. In the case of the sun, the interior of the mass finally reached a high temperature, sufficiently high so that nuclear reactions began to take place, the first of which were the burning of heavy hydrogen and helium-3. The

* We note the claim that the latest Russian moon probe has rocks a billion years older than any previously reported. No details have been given as yet.

contraction continued and a reaction took place in which four hydrogen atoms combined to give an atom of helium, with the generation of a very large amount of energy. According to Urey this stage for a star like the sun should require a period of time of something like 10^{10} years. It is believed that our own is about halfway through this cycle, which certainly does not suggest that any great changes in the sun are to be expected in the near future. Ultimately the star expands and becomes a red giant, and other nuclear reactions take place involving the synthesis of the elements.

The early protoplanet earth likewise condensed from gas and dust and accumulated planetesimals at about 0°C. A brief summary of the events which probably took place in the formation of our solar system is presented in Table 19 in Urey (1952). A high temperature stage occurred reaching approximately 2000°C. During this stage there was loss of gases and volatile substances and a low temperature stage followed with further loss of gases and accumulation of planetesimals. In the final stage the earth heated to approximately 900°C and then cooled to its present temperature. Urey (1963) believes it is unnecessary to assume a completely molten earth at any time to account for the present volcano activity, and that the rate of present volcanic activity may be sufficient to be responsible for the entire amounts of basalts produced in the earth's crust during its existence. The crust appears to have grown steadily in time.

As to the composition of the early atmosphere recent discussions have been given by Fesenkov, and by Urey, both presented in Florkin (1960), and also in Oparin (1938, 1957) and Urey (1952a, b). In brief, the earth's present atmosphere is believed to be very different from the original one, especially in regard to hydrogen, which was formerly most abundant, and oxygen which was formerly minimal. In addition to hydrogen, the early atmosphere included methane and the inert gases. Water vapour, carbon dioxide, methane, sulphides and nitrogen were added by volcanic processes. Large amounts of water finally accumulated to form the early oceans, which have apparently been in existence for a considerable time. Recent estimates from studies of continental drift give the age of the Atlantic ocean as only 200,000,000 years.

The early reducing atmosphere suggested by Haldane (1932) and discussed by Oparin, Urey and others was replaced mainly through the loss of hydrogen and the other reducing substances and its replacement by oxygen following the development of green plants. Urey (in Florkin, 1960) stresses the point made by Haldane (1932) that life probably originated under anaerobic conditions because the fermentative metabolic processes of existing organisms of widely different structure are very similar while the oxidative reactions are very different, thus indicating that the anaerobic metabolism is the more primitive. This argument Urey believed to be more conclusive than any drawn from purely geochemical and cosmochemical studies, though it is now clear that the latter confirm the former.

The reducing nature of the early atmosphere had very important consequences for the synthesis of organic compounds suitable and necessary for the origin of protoplasmic systems; compounds which would not be likely to form or persist under oxidizing conditions. We quote from Urey (1952a, p. 221):

"The earth at the terminus of its formation had an atmosphere of water, hydrogen, ammonia, methane and some hydrogen sulfide. The hydrogen was lost and water was converted to oxygen and hydrogen by photochemical reactions in the high atmosphere: and as hydrogen escaped the atmosphere became oxidizing, ammonia was converted to nitrogen, and methane to carbon dioxide. In the course of this change organic compounds occurred and life evolved."[*]

An extensive study of the primitive atmosphere of the earth has been made by Berkner and Marshall[†] (1965, 1966, 1967) and the levels of oxygen which they describe are associated with a theory of the explosive evolution of animal life in the late Pre-Cambrian and early Cambrian, stimulated by the increased amounts of energy available from aerobic respiration. A second phase of eruptive evolution occurred when sufficient oxygen was available to maintain levels of ozone in the atmosphere which would protect land life from lethal "sunburn", at which

[*] From Urey, H. C., *The Planets*. Copyright, 1952, Yale University Press. With permission.

[†] Berkner and Marshall do not accept the validity of evidences of Pre-Cambrian animal life, stating that the same view is "universally noted by all authors on evolution and historical geology". They quote Kummel (1961) to this effect but fail to note that Kummel, 1970 (pp. 52, 53) refers to the Late Pre-Cambrian fossils reported by Glaessner (1966) and others. Kummel has even reproduced some of the photographs from Glaessner's reports.

time dry land became habitable by plants and animals. The invasion of the land took place in late Silurian and early Devonian times with the rapid exploitation of the new ecologic zones which then became available. We will return to the matter of explosive versus gradual evolution later, and the further implications of the theory of Berkner and Marshall.

In these discussions and reports there seems to be a consensus of opinion that the change from a primitive reducing to a secondary oxidizing atmosphere was dependent upon the development and increasing activity of photosynthetic organisms capable of liberating free oxygen. But this view does not require or imply that green plants were the first organisms on earth, nor even that any kinds of autotrophs were the first forms of life. Rather it is likely that first heterotrophic and later autotrophic anaerobes were the more primitive organisms, to be followed by various kinds of aerobes, both hetero- and autotrophic. Berkner and Marshall (1965, 1966) and others have pointed out the great increase in efficiency to be gained by organisms using oxygen in respiration.

The natural chemical production of organic compounds to be used in the construction of the physical materials of life

Urey (1952a) considered that Oparin's arguments that life originated under anaerobic conditions were very convincing. We quote from his book *The Planets* (1952a, p. 153):*

"Many researches directed toward the origin of life have assumed highly oxidizing conditions and hence start with the very difficult problem of producing compounds of reduced carbon from carbon dioxide without the aid of chlorophyll. It seems to me that these researches have missed the main point, namely that life originated under reducing conditions, and it was only necessary for photosynthetic pigments to become available as free oxygen appeared."

He goes on to say that many anaerobes exist at present which approximate the types of organisms which could exist under such

* With permission, courtesy of Yale University Press. Copyright 1952. From Urey, H. C., *The Planets*.

primitive earth conditions, and thus they are the prototypes of the first living creatures. And to return to Urey (1952a, p. 362):

"It seems to me that experimentation on the production of organic compounds from water and methane in the presence of ultraviolet light of approximately the spectral distribution estimated for sun light would be most profitable. The investigation of possible effects of electric discharges on the reactions should also be tried since electric storms in the reducing atmosphere can be postulated reasonably."

And "profitable" this experimentation did prove to be in the hands of Miller (1953, 1955, 1957a, b). For Miller, using the energy of electric discharges, produced amino, hydroxy, and aliphatic acids from a mixture of methane, ammonia, hydrogen, and water. It is argued that the same compounds would be formed if the earth had a reducing atmosphere but would not be synthesized on earth in an oxidizing environment. Furthermore the energy utilized in their formation was more likely to be obtained from ultraviolet light than from electric discharges, though the same results would be obtained in either case. But the activity of ultraviolet light for such reactions depended on low levels of oxygen and ozone, in other words on the assumed existence of a reducing atmosphere.

Fox (in Florkin, 1960, p. 148) has pointed out that, "the prebiochemical distance from such organic compounds as amino acids to the origin of life, must be quite large". Additional synthetic activities must have produced proteins, nucleic acids, and many other substances. It is to the synthesis of the proteins that Fox has given special attention.

Polymers of various amino acids were experimentally produced by heating them at 200°C for three hours. Positive biuret tests were observed in many instances, indicating that the conditions were suitable for the formation of peptide bonds. This is especially true if mixtures of amino acids are used, and Fox has, on the basis of these results, developed his "Chemical Theory of Spontaneous Generation", as yet incomplete, but suggesting many parallelisms between the pathways of synthesis involved in spontaneous generation on earth, and the anabolic pathways of present-day organisms. According to Fox, "Haeckel's Law" that ontogeny repeats phylogeny is thus obeyed at the biochemical as well as the biological level. The present status of the

B

"biogenetic law" may be clearer after its discussion in later chapters. Suffice it to say here that its general validity as a dependable guide to remote phylogeny is in serious doubt.

The problems of protein synthesis have been discussed also by Akabori (in Florkin, 1960) in his report "On the Origin of the Fore-Protein". It is Akabori's view that the formation of the early proteins did not result from the condensation of available amino acids but rather from the production of aminoacetonitride from formaldehyde, ammonia and hydrogen cyanide followed by its polymerization on a solid surface and the hydrolysis of the polymer to polyglycine and ammonia. Finally the polyglycine received the addition of various side chains and the production of seryl and threonil residues resulted. Further reactions are described which could have resulted in the production of other amino acids and their polymers.

Incomplete as is this account of the synthetic activities which may have resulted in the formation of the first living things, the emphasis throughout has been on natural chemical evolution, rather than on any rare biochemical accidents. Life, we believe, would appear whenever and wherever there was enough "hot dilute soup" (Haldane, 1932, p. 155) and the attendant conditions were suitable. According to Urey (1952) there may have been sufficient organic matter accumulated in the early oceans after 2×10^9 years to have reached a level of one per cent. If the early oceans had contained only one-tenth as much water as at present, the level of organic matter would have approximated a ten per cent concentration. Such amounts of organic matter, whether widely distributed or concentrated in local pools, would be more than adequate to allow the synthesis of the first protoplasms and their maintenance until the capacity for autosynthesis evolved.

The nature of the first living matter

It has been assumed for many decades that since animals and other heterotrophic organisms are now dependent on autotrophic plants, that the first forms of life on earth must have been autotrophs. This view implied that the whole transition from inorganic through organic compounds to living things was one series of events and that no accumulation of non-living organic compounds had occurred in the process. The improbability of such a sequence of rare events is staggering. This would be in fact the most un-natural of all illustrations of un-natural history. Such views have largely given way to the more recent theories, as developed by Oparin and later by various workers cited in our present discussion.

Significant points bearing on the nature of the first life forms have been made by several authors. For Oparin (1957) the first living systems were described as coacervates, capable of assimilating organic nutrients from the environment and, in consequence, of growth and of the crudest possible reproductive fragmentations, which could, however, transmit enough of their systems to some, at least, of the fragments produced to perpetuate the system long enough for natural selection to act. Precise reproductive machinery, on this theory, had not yet evolved. This suggestion is more nearly related to the point of view of Lindegren (1957 a and b), that protoplasm of the nature of cytoplasm preceded the formation of nucleic acid, genes and chromosomes, and this view is favoured by Keosian (1964) as being the more probable. On the other hand, Muller (1955, 1967) is the chief proponent of the theory that the first living thing was a gene, and that this gene developed the capacities to produce cytoplasmic products required for a minimum viable system. Our view is that a set of genes with its organized DNA and its production of RNA is not now sufficient to produce an organism and this conclusion has been strongly supported by Commoner (1962, 1968). It is sometimes overlooked that no organism has ever developed from a spermatozan unless it is associated with an essentially complete cytosome.

As to the auto- or heterotrophic nature of the earliest life, further consideration supports the heterotrophic position. In the first place, under the reducing conditions of the primeval atmosphere, the first life must have been anaerobic and Urey (1952b) refers to the present existence of anaerobes which may serve as the prototypes of the most primitive organisms. Furthermore reducing conditions may have continued even up to 800,000,000 years ago, though the earliest definite plant remains with recognizable structural characters are believed to be about $3\cdot2 \times 10^9$ years of age, from the fig-tree formation of South Africa. These organisms were bacteria and blue–green algae and

Barghoorn (1971) has written an excellent account of the discoveries which he and his colleagues have made during their explorations. There have been several reports as follows: Barghoorn and Tyler (1965), Barghoorn and Schopf (1966), Schopf (1970).

According to Barghoorn (1971), the first organisms were most likely heterotrophs but they had to be followed rather promptly by autotrophs, such as certain bacteria and blue–green algae. These were prokaryotic having no formed nuclei nor organelles such as mitochondria and ribosomes. They were "essentially sexless" lacking the capacities for mitosis and meiosis which are required in normal sexual reproduction. Of these early organisms the blue–green algae at least would produce free oxygen which could begin to accumulate, at least locally, and make aerobic life possible.

The Gunflint formation of Ontario, is about 2×10^9 years old, and it has yielded eight genera and twelve species of primitive plants including bacteria and filamentous and other blue–green algae, probably all autotrophs. There were also some organisms of uncertain nature.

To conclude the discussion of Pre-Cambrian plant life, we turn to the late Pre-Cambrian Bitter Springs formation of Northern Australia. This is dated as approximately one billion years old. Schopf (1970) believed that his fossils represented three bacterium-like species, twenty representatives of blue–green algae, two certain genera of green algae, two possible species of fungi and two forms of doubtful relationships.

The nature of the early plants is better known than that of the early animals, for we have no good fossils of animals earlier than the late Pre-Cambrian. A rich find which includes some relatively advanced types has been described by Glaessner from the Ediacara Hills of South Australia. Among these were jellyfish representing at least six extinct genera, soft corals related to living sea pens, segmented worms and two animals of unknown types, viz., *Parvancorina* and *Tribrachidium*. Rare fossils of similar age and apparently related to the sea pens have been reported from England and South Africa. A fuller discussion and comparison of these various Pre-Cambrian fossils may be found in the reports by Glaessner (1958) and by Glaessner and Daily (1959). In these, comparisons are made between some of the South Australian fossils, especially the sea pens and the similar South African fossils (*Pteridinium* and *Rangea*). Aside from possible sponges indicated only by spicules, no animals of simpler structures than those mentioned have been found among these early fossils. They may then be a considerable distance, in evolutionary terms, from the real first animals, and we are obliged to refer to the simplest existing animals for clues as to the probable natures of their ancestors. We must bear in mind, however, that existing organisms, plant or animal, are the terminal representatives of long lines of ancestry, and however simple they may be, the likelihood of their showing truly primitive structural organizations is small. We are thus forced to look back from the terminal twigs of the evolutionary tree or trees, to try to see what their phylogenetic roots may have been. In doing so, if we realize that the highly specialized processes of mitosis and meiosis and of sexual reproduction are unlikely to be primitive, we may at least make some reasonable guesses as to the probable nature of the most primitive animals. And if we understand that the simpler the existing animals we study, the closer we will be to the roots of animal evolution, and the more likely we are to be right in our view of the beginnings. Let us turn away from the birds and the bees and *Drosophila*, and attend to some of the simpler animals we now *know*. Even these will no doubt be advanced in some respects over their own original ancestors.

It is our hypothesis that these ancestral animals may have been simpler than most existing Protozoa or Metazoa, for these have commonly developed mitotic cell division and thus have the capacity for maintaining a constant relationship between nuclei and cytosomes. On the other hand in our presumed ancestors there were no definite nuclei, nor mitotic spindles nor meiotic behaviour. There were not even any precise relations between the relative amounts of nucleoplasm and cytoplasm because no mechanism existed to keep them in comparable growth, and reproductive, phases. This was the premitotic and presexual period. Growth and reproduction were irregular and in consequence variability was relatively great while heritability was relatively poor. "Evolution was rapid and characterized more by change than by the conservation of change" (Boyden, 1953). In fact evolution may have been too rapid for the existing conditions and this would be true until more perfect genetic machinery had been evolved, which would make it possible to stabilize a successful protoplasmic system. And no doubt natural selection would act then, as it has on many subsequent occasions [Schmalhausen (1949)], to create stability, rather than to promote variability.

It is our belief that it was during this period of primitive life, before the major types of protoplasmic organization were more fully developed and "patented" that major steps in evolution occurred. This was long before fossilization could take place readily, the organisms themselves being devoid of hard parts and relatively few in number.

The present generally accepted view that all organisms are genetically related, of course, implies that life began on earth only once in a "rare accident" (Breder 1942), or that all surviving lines of descent, at least, originated in one organism. But it is very difficult for me to conceive that the conditions which existed at the time of life's origin on earth were of such an extremely local character. In accordance with Oparin's general theory, coacervate droplets could be formed readily and repeatedly under the prevailing conditions and these could be of many compositions even in the same environment. An environment which was favourable for the formation of suitable coacervate droplets would be likely to produce many of these, some differing only slightly, and thus if a stable and yet dynamic system was achieved in one of these, it seems probable to me that other similar but not necessarily identical systems would reach the level of living at about the same time or throughout the period when such conditions prevailed. In other words, polyphyletic origins would be the more probable occurrences under the conditions of natural chemical evolution.

Of course, it is implied in the monophyletic theory of life's origin on earth that from the first living and surviving and reproducing organism all subsequent organisms have descended, and thus the original system must have had practically unlimited capacities for evolutionary change. But once a living protoplasmic system achieves the capacity to transmit and perpetuate such a system, the capacity for change becomes limited! There is certainly nothing in the available knowledge regarding genetic mechanisms today, or in the fossil history of animals for 600,000,000 years to indicate that any surviving protoplasmic systems have unlimited capacities for evolutionary change. Oparin points out that natural selection would begin to act at the earliest stages in favour of the stabilization of types of living systems, while yet allowing evolutionary increases in the efficiency of such systems. To focus attention on the known genetic units themselves, we see in multiple allelic series, that when genes mutate they generally continue to affect the same characters in similar ways, in other words, even genes are limited in their capacity for viable change. It is understandable, then, that since the beginnings of our fossil records, there appear to have been no major types evolved from other major types, such evolution, if it did occur, having taken place before the fixation of the original major phyletic types.

On the view of polyphyletic origins then, such general resemblances in protoplasmic systems as we find in *all* living organisms do not necessarily mean genetic relationship. Rather, such resemblances may be fundamentally convergent, and represent only the minimum requirements or "conditions of existence" which all living systems must possess.

One of the most striking similarities among present-day organisms which would naturally support belief in a strictly monophyletic origin of life, is indirect cell division or mitosis, and its modification for gametogenetic purposes, meiosis. These processes are relatively complex and may have required a very long period of time to bring them to their present perfection in specialized organisms (Boyden, 1953a). However, prior to and during this time there were no such similarities among the organisms which then existed. Furthermore, there is some variability in these processes still (Schrader, 1953) and in Protozoa, for example, a wide range of types of cell division exists. Thus there may be amitosis as well as mitosis; intra or extranuclear spindles; centrioles may be present or absent; meiosis may be accomplished in one or two divisions; and it may be associated with sporulation as well as gamatogenesis. Such variability leaves the door open to the conclusion that there may have been a polyphyletic origin of mitotic cell division. Finally, convergence has occurred in many kinds of organisms as evident in many biochemcial, structural and functional characters. In fact there are few kinds of characters which are not convergent in some groups. Multicellularity, symmetry, number of germ layers, presence of exoskeletons, paired-jointed appendages, wings, gills, sense organs, four-chambered hearts, constant body temperature, parthenogenesis, metagenesis, presence of hemoglobin, all these and many more characters appear to be convergent in some groups. The existence of mitosis and meiosis in organisms as diverse as the Protista, Metazoa, and Metaphyta therefore lends itself to the interpretation that it is another illustration of convergence differing from other such cases in degree rather than in kind.

Pantin (1951) and Grimstone (1959) have reported on "Organic Design", in which the points are made that nature seems to be building some of the elementary and limited structures from "standard parts." For example, the flagellae, so commonly constructed with the "9 plus 2" fibrils seem to suggest a kind of basic patent which must be followed by organisms of many kinds whether their genetic relationship is close or distant or non-existent.

In this connection it seems that the likelihood of convergence in evolution is related to the number of possible alternative expressions which the structural organization is known to have. For instance, in the matter of cellularity, the kinds of alternative expressions are only three, non-cellular, unicellular, and multicellular. Thus if there have been more than one change from one to any other grade of such organizations, convergence necessarily results. Evolutionary transitions from non-cellularity to cellularity are unknown if we deny those mistaken claims that Protozoa, and the zygotes of all Metazoa, are acellular. But the transition from the unicellular to the multicellular state is witnessed in all cases of ontogenetic development in Metazoa where sexual reproduction has taken place, and is believed to have occurred independently in the early evolution of several kinds of plants and animals, evolved from various Protista. The same considerations would apply to metamerism (only two alternative expressions); to jointed and non-jointed appendages (only two alternative expressions) or to symmetry, where the possible expressions are: asymmetry; and spherical, radial or bilateral symmetry together with such combinations as may occur among them.

The evolution of mitosis and meiosis

Under favourable conditions, it would be advantageous for any successful kind of organism to become stabilized to the extent that it could reproduce itself with reasonable accuracy. In fact no such organisms could long survive without something like this capacity. However long it took therefore, natural selection would tend to preserve every effective step toward the improvement of the reproductive machinery. My own estimate of the time required for the evolution of mitosis was of the order of magnitude of "hundreds of millions of years". This estimate was based on the complexity of the typical mitotic process and on the assumption that the rates of metabolic processes and consequently of growth and reproduction in the assumed primitive organisms were much lower than in their modern descendants. Crosby (1955) criticizes this view suggesting that mitosis could have evolved quite rapidly. However, his theory assumes that the first living organisms were groups of primitive genes, which soon acquired the ability to duplicate simultaneously, and thus you would have the first organisms acting essentially as chromosomes, dividing mitotically and it needn't have taken much time at all! This sounds very much to me like a denial of the occurrence of the evolution of mitosis since it was all there to begin with!

As Crosby puts it, "The thesis presented in this paper is that mitosis evolved simultaneously with the first living organisms." Obviously, Crosby's criticism of my estimate of the time required for the evolution of mitosis based on his own theory of the lack of evolution of mitosis is wholly irrelevant.

To return to our theory of the evolution of mitosis, any development which could act to improve the "copying process" (reproduction) in successful organisms under favourable conditions, would have selection value. The development of chromosomes helped in the co-ordination of gene duplication; the development of spindles made possible a more equal transmission of the genetic code to the daughter cells; and the proper timing of nuclear and cytoplasmic divisions made it possible for the daughter organisms and their descendants to maintain a definite size and efficiency. Thus whether the organisms were to become primitive Protozoa or Metazoa, the perfection of the mitotic process was a great advance, for now these evolving organisms could "hold fast to that which was good", whereas before they were able only to "try all things" (Boyden, 1953a). And in the case of the primitive Metazoa as well as the multinucleate Protozoa we would now, for the first time, have the same genetic code in all parts of their bodies, which would be necessary if these organisms were to have the capacity to regenerate any missing part and reproduce asexually from many parts of the body.

Apparently, the most common view among biologists is that Protozoa evolved before Metazoa and that the latter were derived from the former by either a process of colony formation (Hyman, 1940, 1942; Hardy, 1953) or by the cellularization of a multinucleate organism (Dobell, 1911; Hadži, 1953, 1963). Some, however, consider it improbable that the Protozoa are phylogenetically older than the Metazoa. In our view we have compromised to the extent that our Prozoa were believed to be not as advanced as either the modern Protozoa or the Metazoa and probably gave rise to both independently. But in any case the evolution of mitosis probably occurred during the presexual period and hence it was asexual reproduction which did not, strictly speaking evolve into sexual reproduction but rather created the conditions which made sexual reproduction possible and successful.

It is customary in these times to consider that asexual reproduction has no genetic and evolutionary significance, and never has had any. The current definitions of species (Dobzhansky, 1937; Mayr, 1942) are not applicable to asexually reproducing organisms. Neither are they, as a matter of fact, applicable to self-fertilizing organisms. In either case, each individual is reproductively isolated from all others, which is the extreme limit for the isolation of biotypes. Nevertheless such organisms live in populations which have specific characteristics and they also evolve. It is generally thought that asexual reproduction involves totally different kinds of genetic mechanisms than sexual, reproduction and elementary students often get the impression that chromosomes and genes are not concerned in asexual reproduction; that hereditary variation could not take place; and that, therefore, the products of asexual reproduction show little or no variation!

All these misconceptions arise from a lack of perspective in the study of evolution and from too great a concentration on micro-evolution in highly specialized organisms. Better clues in regard to the early stages of evolution should be obtainable if we look at some of the simplest free-living animals now known. Before the evolution of mitosis there could be no regularity to the duplication of genes and chromosomes and no relative constancy of cell size since this would require a synchronization of nuclear and cytoplasmic divisions.

Certainly there could be no regular alternation of haploid–diploid states. Our ancestral Protozoa (Prozoa) therefore may have been like some of the organisms shown in the illustrations of Kudo (1960).

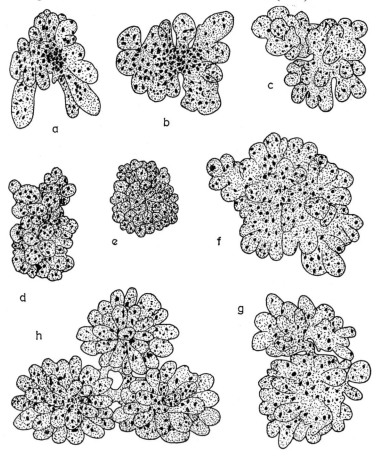

FIG. 2.1. Eight individuals of *Pelomyxa carolinensis*, seen undisturbed in culture dishes, in which mitotic stages occurred as follows (×40), observed by Kudo (1960): a, early prophase; b, c, late prophase; d, metaphase, e, f, carly and late anaphase. g, h, late telophase to resting nuclei: g, plasmotomy into two individuals; h, plasmotomy into three individuals. From Kudo, R. R., *Protozoology*, 4th edition, 2nd printing (1960), Fig. 71, p. 173. Courtesy of Charles C. Thomas, Publisher, and permission of the author.

Our Fig. 2.1 illustrates the process of plasmotomy as observed by him from living material. The individual indicated as *g* completed its division into two individuals in fifteen minutes; and specimen *h* divided into three individuals within the next twenty-two minutes after his drawing was completed. These Pelomyxas could represent truly primitive Prozoa except in regard to their mitotic cell division. In a premotitic state, the size of the nuclei, the size of the organisms, the number of nuclei in each and even the constitution of the nuclei would vary due to the primitive state of their reproductive processes. But if there were a sufficient number of nuclei in each daughter it would possess all the necessary genetic determinants, as Gabriel (1960) has suggested. It is unfortunate that so many students of evolution cannot grasp the idea that the reproductive processes of our primitive organisms were themselves equally primitive.

Other simple reproductive processes are shown in our Fig. 2.2. The budding in *Myxidium lieberkuhni* and the plasmotomy in *Chloromyxum leydigi* and *Sphaeromyxa balbiana* are about as simple as any reproduction processes can be and they still persist! Obviously natural selection passes them on, as it does the reproductive processes of many flagellates but still without benefit of sex. Sex is not essential to life and the views of those such as Dougherty (1955) and Stebbins (1960) that asexual reproduction was derived from some form of sexual process are not convincing.

Stebbins (1960) refers to Sonneborn's excellent report (1957) claiming that it has given "good evidence" that cross-fertilization evolved into self-fertilization and then into asexuality. This is not at all true for Sonneborn was describing the situations as they existed in varieties or species of Paramecium or of the symbiotic flagellates of wood-roaches. These are higher Protozoa and their sexual processes are far removed from any that might be considered primitive. The primitive reproductive machinery was probably defective and even incapable of stabilizing a viable organism—the last thing that was needed was reproductive processes that would increase the already too great variability. Generally speaking the trend in *primitive* evolution cannot be from more complicated and advanced to simpler states. And if asexual organisms have survival capacities along with sexual organisms why was it necessary to go through a sexual phase in order to get them? With

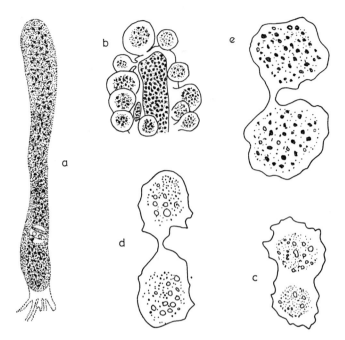

FIG. 2.2. Asexual reproduction in Protozoa. a, b, budding in *Myxidium lieber-kuhni*; c, d, plasmotomy in *Chloromyxum leydigi*; e, plasmotomy in *Sphaeromyxa balbiani*. From Kudo, R. R., *Protozoology*, 4th edition, 2nd printing (1960), Fig. 70, p. 172. Courtesy of Charles C. Thomas, Publisher, and permission of the author.

sufficient perspective we may be able to understand how asexual reproduction could have had a very significant evolutionary role and that:

1. It was probably the primitive and only kind of reproduction which perpetuated life for the long period of time during which genetic mechanisms were evolving into their modern perfected states.

2. Its chief role in evolution was to aid in the stabilization of the early states of protoplasmic organization.

3. This stabilization was accomplished finally through the evolution of mitosis, which made possible a true reproductive process giving

similar daughter cells among unicellular forms and making it possible for simple many-celled forms to regenerate or bud from many parts of the body.

4. This stabilization process did not prevent all variability but allowed for types of polymorphism as in certain Protozoa, Coelenterata, and Platyhelminthes even today and always permitted hereditary variations by mutation.

5. Finally, without asexual reproduction it is highly unlikely that there would have been any evolution of sexual reproduction because the latter can only be biologically successful with complex and highly specialized cells and processes.

Now if the evolution of mitosis required a long period of time, what about meiosis? The transformation of mitosis into meiosis would seem to involve relatively simple changes compared with the probably polyphyletic evolution of mitosis out of amitosis. Thus the main differences between mitosis and typical two-division meiosis are two: (a) Synapsis of homologous chromosomes in preparation for the first meiotic division and (b) failure of the chromosomes and genes to duplicate during the second meiotic division. These changes are believed by Darlington (1958) to have come about at one step. It was his view that meiosis was an abrupt change which permitted no halfway house. The chromosomes must either have reduced or failed to reduce if they were to keep their genetic character. And again Darlington believed that the origin of meiosis and sexual reproduction represented the most violent discontinuity in the whole of evolution, which demanded not merely a sudden change but a revolution.

We are not sure that the change from mitosis to meiosis was as sudden as Darlington states. Cleveland (1949, 1950, 1951) has described many variations and permutations of mitotic and meiotic processes. Instead of the common two-division meiosis there may be a one-division meiosis. And, therefore, the typical cycle with tetrad formation, and genetic segregation associated with crossing-over, may be lacking. It is apparent also that meiosis has evolved in connection with sexual cycles and also in cycles of sporulation and that these were independent sequences. But in all events the evolution of meiosis from mitosis could have been relatively easy and of short time requirement and probably occurred independently among different phyletic lines.

The Age of Stabilization

The last 600,000,000 years show no revolutionary changes in the apparent mechanisms of evolution or in their products. All the major types of animals likely to leave fossils are represented in the early Paleozoic rocks. This period of evolution is therefore characterized as much by the conservation of change as by the changes themselves. It is evident that in the processes of establishing the major animal types, limits have been built into protoplasmic systems beyond which living patents cannot be extended. Of evolution by gradual variation leading to adaptive radiation there has been a great deal and in the Insecta for example almost infinite variability on a basic patent has taken place. But so far as is known over this long period of time no insect and no arthropod has evolved any different pattern leading to another phylum. The evidences relating to the Age of Stabilization do lend themselves to interpretation on the Neo-Darwinian basis, all this evolution being subspeciational and completely graded in origin, and the gaps between species having been secondarily acquired. But such evidences do not necessitate the conclusion that *all* evolution has been so gradual.

Indeed, there are alternative views. Clark (1930), Goldsmchidt (1940), and others have believed in large evolutionary steps producing species, genera, or even higher categories at single steps. Willis' (1922) study of the relative frequencies of monotypic and polytypic genera led him to conclude that many of the evolutionary steps were of generic magnitude to begin with. Hardy in a refreshing report entitled "Escape from Specialization" (Huxley, Hardy and Ford, 1954) calls attention to the possible sudden evolutionary changes of considerable magnitude which may have been associated with paedomorphosis, a subject which De Beer has emphasized in recent publications (1958, 1959). But I would like to point out that in none of these cases has there been any real "escape from specialization" even if De Beer's views are correct— rather there has only been a substitution of one kind of specialization for another one. . . . Furthermore, such instances of paedomorphosis as may have occurred, probably took place near the beginning of this age of stabilization, or even earlier, and therefore we can believe that the last 600,000,000 years is characteristically an age of evolution in which there was both the conservation of changes and the conservation of evolutionary mechanisms as well.

Let us return to the consideration of the whole of evolutionary history from the point of view of comparative mechanisms. It is possible to show the major events and their sequence in chart form.

THE GREAT AGES OF EVOLUTION
Time in Millions of Years

Present	Era	Characteristics of Eras	Nature of Evolution
	Cenozoic	Many phyla	Age of relative
	Mesozoic		STABILIZATION
	Paleozoic	Fossils first abundant	Adaptive Radiation
600			
	Pre-Cambrian	Fossils rare and uncertain	Reproduction More Perfect: Mostly Sexual
		Heterotrophs and Autotrophs	Mitosis and Meiosis Age of ERUPTIVE EVOLUTION
		Oldest Sedimentary Rocks	Reproduction Imperfect Asexual and Premitotic
3500		Heterotrophs Origins of Life	Age of POLYPHYLETIC ORIGINS
		Accumulation of organic compounds	Age of NATURAL CHEMICAL
4500		Sterile Earth	EVOLUTION

FIG. 2.3. The great ages of evolution. The figure attempts to show the characteristics of the various eras of geological time and especially the relative durations and the natures of the evolutionary changes occurring in them. Modified from Boyden (1953). Courtesy of *Evolution*.

Our Fig. 2.3 is modified from that shown previously (Boyden, 1953).* The modification has consisted only in moving back the dates to bring them into accord with the current estimates. As to the sequence of events, our chart is essentially the same as that shown by Barghoorn (1971), which shows the ages as the current estimates indicate.

Barghoorn emphasizes the probable importance of the evolution of the eukaryotic condition and the subsequent development of sexual reproduction. We quote (pp. 40 f.):

"An oxygen-poor environment was certainly a major obstacle in the path of oxygen-dependent heterotrophs. In addition to the evidence provided by oxidized Pre-Cambrian formations, however, a number of recent studies indicate that the earth's atmosphere had actually accumulated enough oxygen to establish some kind of ozone shield considerably before the end of the Pre-Cambrian. The appearance of eukaryotic organisms late in the Pre-Cambrian, as indicated by the green-algae of the Bitter Springs formation, provides a better explanation for the failure of higher organisms to appear until an even later time. The fundamental key to evolutionary progress is genetic variability. Sexual reproduction, which involves the re-combination of heritable characteristics, is the highway to genetic variability and all its consequences, including the increased complexity of form and function at all levels of organization, that are thereafter apparent in the course of evolution."†

Barghoorn makes no mention of the possibilities of eruptive evolution due to the imperfections of the primitive reproductive mechanisms of premitotic organisms.

Summary

1. Current evolutionary theory suffers from a lack of perspective in that the well-known genetic mechanisms of highly specialized organisms have been assumed to have existed from the very beginnings of life on earth and known simpler genetic mechanisms have been completely ignored.

* Courtesy of *Evolution*, 1953.
† From *The Oldest Fossils* by Elso S. Barghoorn. Copyright by Scientific American, Inc. All rights reserved. With permission.

2. The assumption is still frequently made that autotrophic organisms must necessarily have preceded heterotrophs because animals are now dependent on autosynthetic organisms. Oparin and others have indicated that there is no such necessity, nor even for the assumption of a strictly monophyletic origin for all of the surviving protoplasmic sysytems.

3. The simplest free-living animals are rhizopods and these should once again be considered as the prototypes of the first animals. Models of such remote ancestors are modern Pelomyxas which, however, would originally have been in a premitotic and presexual state.

4. There are many known connecting links between Rhizopoda and Flagellata which are usually interpreted as evidences for the evolution of Rhizopoda from Flagellata. But these evidences lend themselves just as well to the contrary conclusion, viz., that Flagellata evolved from the simpler Rhizopoda, which is in fact more in accord with the general nature of primitive evolution from simpler to more advanced states.

5. There is no necessity of assuming that the most primitive Protozoa were uninucleate and haploid, and it is difficult to believe that the machinery was available in the beginning for the maintenance of such an organization. A simpler solution to the problem of early living would be that of a multinucleate state, with varying numbers of nuclei and states of ploidy.

6. It is assumed that mitosis with its complex and precise machinery of duplication of genes, chromosomes and often of cytosomes, was derived from the much less precise and effective processes of amitosis.

7. It is assumed that mitosis evolved in primitive asexually reproducing organisms, aided by natural selection, by which it was favoured because as the precision of reproductive processes increased, the stabilization of the organisms was accomplished. Furthermore, these organisms could now for the first time, better regenerate and asexually reproduce from many parts of their bodies, since mitosis now provided the same genetic code in all of their cells.

8. The changes necessary to evolve meiosis from mitosis are relatively few and they probably occurred on many occasions. From then on the organisms could for the first time maintain regular haploid–diploid alternative states and they were therefore prepared to exploit and reap

the advantages of sexual reproduction. Up to this time effective recombination mechanisms in animals did not exist.

9. With these considerations we are able to view evolution with greater perspective, and to see it as comprising several great ages. First the age of natural chemical evolution during which organic materials were synthesized and accumulated. Next the appearance of the simplest possible living systems, viz., heterotrophs, probably polyphyletically. Then as the organic nutrients became depleted, the addition of autotrophic organisms, which, ever after, were necessary in the balance of life. Finally, the age of relative stabilization, which includes the last 600,000,000 years or more, during which the basic patents of animal structure have been maintained but never infringed. But these basic patents must have been first established long before the age of stabilization and probably by more primitive evolutionary and genetic mechanisms than those characteristic of the specialized higher animals of today.

As to the duration of the several ages we have made some guesses. Oparin speaks of several hundred million years as the period during which a great variety of organic compounds was accumulated. This would seem to be an under- rather than an over-estimate. The length of the Age of Polyphyletic Origins of the anaerobic heterotrophs is quite indeterminate as is also the matter of the polyphyletic origins themselves! However, it seems unlikely that there were such extremely localized conditions of time and space as would lead to a strictly monophyletic origin of life; or that a successful living system could spread and extinguish all other such systems before the latter could become locally established. But in regard to the Age of Eruptive Evolution we have some basis for placing its beginning at about three billion years ago. In the first place, the complexity of Cambrian and of the relatively few Pre-Cambrian fossils, is so great that, even assuming rapid evolution, a very long period of time would be required for their development. Oparin (1957) places the appearance of oxygen sufficient for aerobic respiration at about 700 million years ago. However, this seems too short a time for the evolution of the respiratory adaptations of these Cambrian animals. Either the respiratory adaptations were evolved under anaerobic conditions (which seems unlikely), or there was a longer period of time for aerobic life.

We believe that some much needed perspective has thus been added to the study of evolution and that this perspective can be helpful in our further study of the primitive evolution of the lower Metazoa.

Tentative and even unprovable as many of the assumptions regarding the long view of evolution on earth must be at present, they yet have the virtue of showing the short-sightedness of many of the current aspects of evolutionary theory, and the need of further study of these problems. The next chapter will attempt to illustrate some of the ways in which greater understanding may be obtainable through this added perspective.

Urey (in Florkin, 1960) considers that the enormous iron deposit of the Pre-Cambrian, some 2×10^9 years ago, may mark the time of transition from a reducing to an oxidizing atmosphere.

Vinogradov (in Florkin, 1960) reviews evidence for the presence of algae in the Proterozoic and Archean limestones in several geological formations and concludes (p. 26), "Thus we have direct evidence that at least as long as about 2×10^9 years ago, O_2 was present in the atmosphere."

It was during this age also that a variety of autotrophic systems became established, modifying, but not displacing, the already built-in systems of heterotrophic and anaerobic metabolism. Finally, the Age of Stabilization is at least 600 million years old and may be given longer estimates as the discovery of older fossils belonging to already known phyla takes place.

We must return to the extensive report of Berkner and Marshall (1965) which presents views differing in important respects from those of Urey, Vinogradov, and others. In fact, Berkner and Marshall have presented a full account of the production of the earth's atmosphere and its presumed effects upon the evolution of life.

The major points in the theory of Berkner and Marshall may be briefly summarized as follows:

1. The early earth was without a primordial atmosphere, and its secondary atmosphere was derived from local heating and volcanic action.

2. The primitive oxygen in the atmosphere was derived from the photochemical dissociation of water vapour, but the concentration of atmospheric oxygen could not exceed 0·001 of the present atmospheric level because of the Urey self-regulation of this process.

3. The rise of oxygen from primitive levels can only be due to photosynthetic activity, the amount of which depends upon the prevailing geographic and ecologic conditions. During the Pre-Cambrian, ultraviolet radiation would penetrate 5 to 10 meters of water. This would limit life to benthic organisms in shallow pools, small lakes or protected shallow seas, where the environmental conditions would prevent convection which would bring such organisms to the surface and destructive irradiation. Life could not exist in the oceans generally, and pelagic organisms could not survive. But photosynthetic benthic organisms receive enough light at the protected depths to produce free oxygen.

4. When the oxygen so produced reached the value of 0·01 of the present atmospheric level, the ocean surfaces would be sufficiently protected to permit the widespread extension of life to the entire hydrosphere. This oxygenic level is specified as the "first critical level" and is identified with the sudden explosion of life at the beginning of the Cambrian period (600 million years ago).

5. When the oxygen reached 0·1 of the present atmospheric level, the land surfaces were protected from lethal ultraviolet radiation to permit the spread of life to dry land. This "second critical level" is identified with the end of the Silurian period (420 million years ago).

The theory of Berkner and Marshall has many important implications for evolutionary theory. It assumes a relatively sudden explosion of living types in the early Cambrian period. They insist that the fossil record should be read as it is observed—that is, with the sudden appearance of many types of organisms *in the Cambrian*. No long pre-history of soft-bodied, non-fossil producing organisms should be assumed. Unfortunately Berkner and Marshall ignore the Pre-Cambrian fossils described by Glaessner and his associates and others reported from Australia, South Africa, Siberia and England. It is true these latter fossils are apparently late Pre-Cambrian, and only a few tens of millions of years may be at issue. Let us not therefore minimize the important implications for evolutionary theory which derive from the work of Berkner and Marshall.

Perhaps the most significant of these is that benthic organisms would be much more likely to evolve early than pelagic organisms. This has a definite bearing on our own theory of the evolution of the lower Metazoa and would support us in the conclusion that hydroids are more primitive than jellyfish.

Another implication is that there was probably some neoteny involved in the evolution of primitive Metazoa, since marked morphological changes may occur where neoteny takes place. The same consequences would be associated with paedomorphosis (De Beer, 1959).

CHAPTER 3

PROPHYLOGENY:
PRIMITIVE EVOLUTION IN THE
EUMETAZOA

A new approach to these old problems

THE ORIGIN of the phyla included in the lower Metazoa has been the subject of much study and controversy for decades and is still under active discussion. Hyman (1940) refers to a number of the theories and inclines to the view that Metazoa arose from some kind of flagellated colony in which there occurred first a differentiation into somatic and reproductive cells and then a differentiation into locomotor-perceptive and nutritive types, through the wandering of the latter into the interior. The parenchymula or stereogastrula thus formed may represent the common ancestor of the Cnidaria and Platyhelminthes.

Other recent discussions bearing on these two phyla may be found in the papers of Greenberg (1959), Hadži (1953), Hand (1959), Hanson (1958) and Jägersten (1955 and 1959) which present many differences of opinion in regard to the relative primitiveness of cnidarians and flatworms and as to which type was ancestral to the other. Hyman (1959) has recently presented her own definite views in the Retrospect to vol. 5 of her series.

For a remarkably clear analysis and summary of the facts and theories bearing on the origin of the Metazoa and the relative primitiveness of the lower Metazoan phyla, the book of Kerkut (1960, chaps. 5, 6) should be consulted. There is also the book by Jovan Hadži (1963) which presents a large part of his life's work.

We shall attempt here a brief summary of the essential points listed in Hadži's theory of the evolution of the Metazoa.

Animal evolution has proceeded in the sequence Flagellata → Ciliata → Acoela and other Bilateria → Anthozoa → Scyphozoa → Hydrozoa. The chief bases for this interpretation are the following:

1. Bilateral symmetry was original because the stem forms were motile.

2. This bilateral symmetry was transmitted from the Protozoa to the Acoela and thence to the bilateral Anthozoa from which the other classes of Cnidaria evolved, losing their bilateral symmetry on the way because of the adoption of a partly sessile life.

3. Polyps were primitive and medusae were later forms in the Scyphozoa and the Hydrozoa.

4. The hypogenetic Cnidaria (e.g. *Trachylina*) are derived forms and not primitive.

5. The Cnidaria are not really "diploblastic" and there is a greater similarity between Turbellaria and Cnidaria than the claim that the former are triploblastic and the latter are diploblastic would indicate.

Now we are glad to say that some of the parts of Hadži's theory are plausible and acceptable. In particular we are in agreement with regard to the following matters:

1. The evolution of Metazoa from Protozoa was more likely to occur by cellularization than by colony formation, but not necessarily from the Ciliata.

2. Metagenesis is a valid and significant kind of life-cycle in two classes of Cnidaria.

3. Polyps better represent the primitive form of Cnidaria than do medusae.

4. Such hypogenetic life-cycles as occur in Trachylina are probably secondary simplifications, only more extreme than occur in those Hydrozoa where the free medusae have been reduced to sessile gonophores.

5. Polyps are not "larvae".

Our differences are mainly in regard to the following points:

Hadži's great emphasis on bilateral symmetry as a guide to phylogeny is not warranted. In his own words, "It is certain that a symmetry can be changed much more easily than the whole structure." We find that the quotation is acceptable, for the evidences that symmetry can change within phyla are valid. Furthermore there is no strong or necessary correlation between motility and bilateral symmetry, nor between a sessile life and radial symmetry. Within the Cnidaria even, sessile polyps may be radial, as in Hydrozoa and Scyphozoa, or bilateral as in Anthozoa. And the most motile of all Cnidarian forms are the radial medusae. As to the structural correspondences between Turbellaria and Anthozoa, it is the view of Pantin (1960, 1966), who has studied the structures carefully, that the structural differences are too extensive to permit of an evolutionary derivation of Anthozoa from Acoela, or even the reverse.

We shall return to these matters at the end of this chapter.

Then, in 1963, the book *The Lower Metazoa, Comparative Biology and Phylogeny*, ed. by E. C. Dougherty and others, was published.*

This comprehensive work provides excellent reports bearing upon the phylogeny of the lower Metazoa by most of the principle investigators concerned. Though no report by Hadži is included, essentially similar views were presented by Steinböek who developed them independently. In any event Hadži's own book is now available so the ciliate-acoel theory of the phylogeny of the lower Metazoa is adequately treated.

Though it is impossible to present any real substitute for the reports as given, we can correctly conclude that much diversity of opinion still exists in regard to the relationships which exist between the Cnidaria and the Platyhelminthes and in regard to which was ancestral and which descendent. Hanson searches for homological identities between Ciliata and Acoela and admits that in no case are the major criteria of Remane (1955, 1956) completely fulfilled. It is claimed, however, that in four instances the data come close to meeting the requirements of the criterion. These instances include (1) the nuclei, (2) the contractile organelles, (3) the secretory extrusible elements, and (4) the processes of the sexual phenomena. On the other hand Remane (1963), whose criteria of homology Hanson attempts to satisfy, states that the similarities between the ciliates and aceles are analogous not homologous. "No criterion of homology is given" (p. 27). Remane concludes that the flagellate derivation of the Metazoa is better founded than the theory of ciliate derivation. And so the reports of many other investigators follow with no hint of a consensus of opinion and with evident strong

* This book had its origins in the Second Annual Symposium on Comparative Biology of the Kaiser Foundation Research Institute held in 1960, largely supported by the National Science Foundation, the contents of which were somewhat altered and enlarged and published with the aid of a generous donation from the Cocos Foundation, Inc., Indianapolis, Indiana.

divergence in phylogenetic interpretations. And this remains true even after the publication, under the editorship of Rees (1966), of *The Cnidaria and their Evolution*.

There would be no point in examining these matters again unless a new approach is taken, and this we intend to do from the standpoint of our previous discussion of the probable nature of primitive evolution. There is need of a new principle in phylogeny, and that is the principle that *the primitive representatives of the older phyla evolved by primitive mechanisms*. Where should we look for the present-day clues as to how this may have occurred? Certainly not to the birds and the bees and *Drosophila* where all is specialized including their evolutionary mechanisms! This would be like trying to locate the roots of a tree by sighting back from the tips of its highest brances, a procedure which is not likely to bring clear vision and may even lead to the roots of another tree! Instead, we will attend to the simpler surviving members of these lower phyla and this would be more like sighting from those branches closest to the roots. Furthermore, in these lower organisms we will attend more particularly to their simpler reproductive processes which would be expected to be more representative of the original evolutionary mechanisms. This new approach is in accord with the operating principles of what we have called "Prophylogeny" (Boyden and Shelswell, 1959) and it is to be contrasted with the current methods and discussions which are based mainly on specialized organisms and the specialized mechanisms of all organisms, simple or complex.

Comparison of the chief kinds of reproduction

Reproduction is the process of producing offspring. The word itself implies a periodic process and a copying process. These processes are of two chief kinds: asexual, without gamete formation (agamogony*) and sexual (gamogony), with gamete formation.† In each case there is the transmission from parent to offspring of organized protoplasm with its innate capacities. Under asexual reproduction there are two chief

---

* The terms agamogony and gamogony are from Hartmann (1956).

† An excellent discussion of the distribution, nature, and classification of reproductive processes is to be found in Vorontsova and Liosner (1960).

subdivisions, viz.: fission and budding. The difference between these is not so much a matter of the relative sizes of parent and offspring as a matter of the division of the parental organization in fission contrasted with the production of new growth in budding (Kramp, 1943–4).

Fission may be an irregular process as in laceration, or a regular process without obvious signs of degeneration. In either case it may be binary or multiple; transverse, oblique, or longitudinal; and the fission planes may have been prepared in advance (paratomy) or not (architomy).

Budding may be external as in Cnidaria, or internal as when gemmules are formed in sponges, or statoblasts in Ectoprocta. External buds may detach themselves, as in *Hydra*, or remain attached to form a continuous colony. In the latter case diverse types of individuals may be formed and the condition of polymorphism thus arises. This kind of organic differentiation among individuals with the same genetic code is more characteristic of budding than of fission, in keeping with the greater amount of new protoplasm concerned in the former process.

Sexual reproduction always involves gamete formation, but it may be uniparental and unisexual as in parthenogenesis; uniparental and bisexual as in self-fertilization; or biparental and bisexual as in cross-fertilization. It therefore may, or may not, be associated with recombination, and in fact there is no single and uniform kind of consequence of sexual reproduction. The genetic consequences of sexual reproduction depend on the breeding system. In self-fertilization or even in cross-fertilization with close inbreeding, the production of new combinations of old qualities may be reduced to negligible proportions. On the other hand, there may be greater variability with cross-fertilization but this is not guaranteed, for special genetic devices, balanced lethals for example, may prevent it. It is unwise, therefore, to attempt to make the distinction between sexual and asexual reproduction on the grounds of the union of diverse gametes or on the supposed universal results of such unions. On the contrary the fundamental distinction between them is related to the actual machinery used for the transmission of the genetic code from parent or parents to the progeny. If this machinery includes gametes and gametogenesis we have sexual reproduction, otherwise not, and both the nature and the evolution of these kinds of reproduction show them to be distinctly different phenomena.

A comparison of some of the characteristics of these kinds of reproduction is shown in Table 3.1.*

TABLE 3.1. A COMPARISON OF THE TWO CHIEF KINDS OF REPRODUCTION

Characteristic compared	Kind of reproduction	
	Asexual	Sexual
Simplicity	greater	less
Occurrence	chiefly in lower phyla	in all phyla
Primitiveness	greater	less
Gamete formation	never	always
Meiosis	seldom [1]	usually
Segregation and independent assortment	seldom [1]	usually
Strongly correlated with regenerative capacity	yes	no
Usually requires favourable conditions for growth	yes	no
Role in metagenesis	associated with polymorphism and fixation of the colony to substratum or host	associated with dispersal and production of new colony
Role in evolution	concerned in primitive evolution; reproductive isolation unavoidable	not concerned in primitive evolution; reproductive isolation secondarily acquired

[1] Where meiosis is associated with sporulation.

The nature and distribution of asexual reproduction

The distribution of asexual reproduction in the animal kingdom† is mainly what would be expected of a primitive process. Thus it is a significant and frequent occurrence in Protozoa, Parazoa, Cnidaria, Platyhelminthes, Nemertinea or Rhynchocoela, Entoprocta, Ectoprocta, Phoronida, Annelida, Echinodermata and Hemichordata. Except for the last two taxa one may plausibly interpret the possession

* From Boyden (1954a). Courtesy of *Systematic Zoology.*
† See Vorontsova and Liosner (1960) for many details.

of asexual reproduction as a direct heritage from the past. The lack of any frequent use of asexual reproduction may be as instructive as its presence. Thus we would list the Ctenophora, Aschelminthes, Brachiopoda, Mollusca, and Arthropoda, together with several of the small phyla, as being relatively specialized in development and in structure, a specialization which has brought with it a distinctly lowered regenerative capacity as well as a sole reliance on sexual reproduction.

Relatively little more is required of asexual reproduction than the capacity for growth itself and, in general, asexual reproduction occurs most rapidly under conditions where growth is rapid.* This is not usually the case in sexual reproduction, which more often occurs when living conditions shift toward a less favourable state. Reproduction has been characterized as "growth beyond the limits of the individual" (Boyden, 1954a)† but this characterization surely is more applicable to asexual reproduction, and especially to budding, than to sexual reproduction. The distinctive thing about sexual reproduction, as we have pointed out, is the elaborate and precise series of events preparing for and leading to the formation of gametes and much more than growth is involved on the part of the parent and the developing germ cells.

It is true that in the case of oogenesis there is a growth period which results in the production of the primary oocytes, and it is also true that the retention of this larger size is made possible for the secondary oocyte, the ootid, and ovum, by the unequal cytoplasmic divisions which result in polar body formation. But in spite of all this, sexual reproduction could not adequately be described merely as "growth beyond the limits of the individual".

The high degree of correlation between asexual reproduction and regeneration has often been referred to and no doubt this correlation indicates that the basic machinery and requirements for the two processes are very similar. As pointed out previously, following the evolution of mitosis, each cell of the soma would be provided with the whole gene code and hence would be equipped to transmit the whole code to daughter individuals or to participate in the regeneration of injured or missing parts. Processes of chromatin diminution whereby

* Laceration may occur under unfavourable conditions.
† This characterization was based on the statement of Darwin (1868), "Inheritance must be looked at merely as a form of growth."

only the cells in the germ line retain the whole code and the soma gets less, as in certain nematodes, are apparently later specializations. In such organisms both the capacity for asexual reproduction and for regeneration are lost. Indirect support for the requirement that a regular mitosis is necessary in Metazoan organisms that can reproduce asexually and regenerate well would be provided by a comparative study of somatic mitosis in the lower phyla and in *Arthropoda* or other specialized phyla. In these latter there are apparently many normal and abnormal modifications of somatic mitosis.

Asexual reproduction and metagenesis

The life cycles of many Cnidaria and Platyhelminthes involve an "alternation of generations", that is, a generation sexually produced is followed by one or more generations asexually produced. This kind of life cycle is well illustrated by many Hydrozoa, Scyphozoa, Trematoda and Cestoda, and this is now what is generally referred to by the term metagenesis.

The significance of metagenesis is still the subject of a great diversity of opinion. Hyman (1959) views it as a senseless idea. According to her, the life cycle of an animal must be continuous from the egg to the sexually mature adult. But it should not be overlooked that in the life cycles which exhibit metagenesis, not all individuals arise from eggs, nor do they necessarily culminate in sexually reproducing adults. The apparent source of the confusion seems to lie in the understanding of what an *individual* animal is, and what are *larval* and *adult* stages in a life cycle. In order to make an attempt at a fair appraisal of the significance of metagenesis we need now to define these terms in such a way that we will know when we are dealing with the conditions or stages to which each term properly refers.

Individual. The individual is a physically separate and essentially complete organism; or if a part of a continuous colony, the individuals are morphologically distinguishable and comparable to the typical independent organisms of the same or closely related species.

Adult. The adult is the mature and terminal stage in the ontogeny of any individual organism. Reproduction, either asexual or sexual is common, though not universal.

Larva. A larva is a post-embryonic and typically active stage in ontogeny which usually metamorphoses into the adult stage. Reproduction in this stage (as in cases of neoteny) is relatively rare.

With these definitions in mind we may be able to clarify the nature of metagenesis and assess its biological significance. One of the classical failures in understanding metagenesis is that of T. H. Huxley (1894), who completely misunderstood it. In a Friday Evening Discourse delivered to the Royal Institution (April 30, 1892) the title of which was "Upon Animal Individuality" Huxley is reported to have said:

"There is no such thing as a true case of 'alternation of generations' in the animal kingdom; there is only an alternation of true generation with the totally distinct process of gemmation or fission.

"It is indeed maintained that the latter processes are equivalent to the former: that the result of gemmation as much constitutes an individual as the result of true generation; but in that case the tentacles of a *Hydra*, the gemmiferous tube of a *Salpa*, nay, the legs of a Centipede or Lobster, must be called individuals."

It is difficult to understand how a great man could be so confused as to be unable to distinguish between the production of tentacles on a *Hydra* and the production of other Hydras by budding. Or how Huxley distinguished between "true generation" and untrue generation ("gemmation") when they both produced identical results. The clue was found in his definition of "the individual" in the same discourse.

"The individual animal is the sum of the phenomena presented by a single life; in other words, it is all those animal forms which proceed from a single egg taken together."

The implications of Huxley's philosophical definition of "the individual" are obvious.

1. There is only one kind of reproduction or "true generation"; which is sexual.

2. Only those organisms which were produced sexually can be called individuals; those produced by budding which may even be indistinguishable from those sexually produced, as in *Hydra*, are not individuals. What Huxley would call them is not clear.

3. In a life-cycle such as that of *Obelia* the "individual" includes not only all the budded and attached hydrozooids and gonozooids, but also *all the detached medusae* derived from that colony!

It seems to me that no more useless and misleading definition of "the individual" could have been devised, for in organisms which have the capacity for asexual reproduction no one without an actual pedigree would know whether he was dealing with an individual or not! And it is on these false premises that Huxley concluded that "there is no such thing as a true case of 'alternation of generations' in the animal kingdom"!

The careful appraisal of Kramp (1943–4) is much more helpful, though he discusses the phenomenon in the most general sense and his remarks are valid only in reference to that sense, which is derived in the first instance from the Preface of Steenstrup's report (1845). To quote Steenstrup, "The subject of this essay is the remarkable and till now inexplicable phenomenon in the animal kingdom, that some animals produce an offspring which at no time resembles its parent, whereas, on the other hand, the next generation, or the brood of the offspring, returns to the form of the original parent."

Later on, Steenstrup strongly emphasized the different kinds of propagation in the alternation of generations; and it is in this restricted sense that many zoologists, including ourselves, now use the term. Obviously, Steenstrup was in error when he stated that the offspring "at no time resembles its parent" but in spite of this he had, in my opinion, a better understanding of the true nature of the phenomenon than Huxley or Hyman. A further quotation from Steenstrup will reveal some of the confusion with which he had to contend and the relative clarity of his own ideas.

"In order to prevent my being further misunderstood, as I have repeatedly been in my oral exposition of this subject and in the lately published Danish version of the work, I must remark that the phenomena upon which my view of the alternation of generations rests, are, as every naturalist knows, new only in part, but that they have in this work received another and, in my opinion, a natural explanation. They have generally been looked upon as instances of metamorphoses or transformation, the essential objection being overlooked, that a metamorphosis can only imply changes which occur in the same individual; but when from it other individuals originate, something more than a metamorphosis is concerned. Thus it is quite erroneous to term a *Scyphistoma* the larval condition of *Medusa Aurita*, since a *Scyphistoma* never becomes a Medusa, but is the QUASI mother of some scores of them.

"There is no transition from a metamorphosis to an alternation of generations, and a metamorphosis once commenced, cannot be continued beyond the generation, nor from the living or dead individual to another individual."

Now the proper distinction between a reproductive process and a metamorphosis is the matter of chief importance in understanding metagenesis and this distinction in turn depends on proper definitions for the individual, the adult, and the larva. With the definitions we have already given, the recognition of metagenesis where it occurs should naturally follow. On this basis we interpret the life-cycle of Obelia as truly metagenetic. It is composed of three kinds of terminal or adult stages, viz., the hydrozooids, gonozooids, and medusae, which reproduce but do not metamorphose; and the larval stage, the planula, which does not reproduce but does transform. A Scyphozoan such as *Aurelia* includes adult scyphistoma and medusa stages, and larval stages, viz., the planula and the ephyra. Comparable stages occur in the Trematoda and Cestoda, which will be discussed further later.

It is one of the virtues of Kramp's report that he gives details of several life-cycles where there may be apparent transitions between metagenesis and metamorphosis, or between life-cycles with and without metagenesis. Thus in the monodisc strobila of a Scyphozoan the medusa can be looked upon as a metamorphosed distal part of the scyphistoma or polyp. If after the medusa detaches the pedicel degenerates the process may then be considered nothing but a metamorphosis. But as a rule the pedicel "regenerates" and produces more ephyrulae and the cycle then includes both reproduction and metagenesis. Other instances are given where the distinction is sometimes uncertain. In our opinion, these occasional apparent or real transitions between cycles with and without metagenesis should not be taken to deny the occurrence of metagenesis where there is definitely an alternation between a generation sexually produced and one or more generations asexually produced. In certain of the Cnidaria, at least,

C

such cycles are perfectly normal and regular, and Kramp's general conclusion based on all the organisms that show metagenesis, including Protozoa, Coelenterata, Platyhelminthes, Annelida, and Tunicata, that it is a varied and polyphyletic process, should not be used to discredit it within the Cnidaria, where it is a more uniform process which we believe has a bearing on the phylogeny of this phylum.

Asexual reproduction and phylogeny

Any metagenetic life-cycle involves a succession of two or more kinds of adults, one of which is produced by sexual reproduction, and the other, or other kinds, are produced by asexual reproduction. The question arose early and is still under active discussion in regard to the relative primitiveness of these various types of adult organisms. Briefly stated, the principal contenders for the honour of representing the ancestral Hydrozoan Cnidarian are the polyps and the medusae. The history of the development of this century-old controversy is recounted in many places (Kramp, 1943–4; Rees, 1957, 1966) and the evidences are critically presented in Kerkut (1960).

There would be no justification for a further sifting of the evidences if there were no new facts or principles to be used. The facts, indeed, are there in abundance, but their selection and weighting depends on the underlying phylogenetic principles to be used. Our contribution, whatever its final significance may be, is based chiefly on the following innovations:

1. A better and sharper conception as to the meaning of the terms larva, adult, and individual, so that we need not misinterpret the life-cycles, and

2. A focussing of attention on the simpler and presumably more primitive genetic and evolutionary mechanisms which are still used by three organisms.

These considerations, we hope, will serve to sensitize the student to the facts bearing on the problems of the origin and relationships of the Cnidaria and show him how the new theory of prophylogeny can be applied to such a problem. We have spoken of "prophylogeny" as the study of primitive evolutionary stages and their probable evolution by primitive evolutionary mechanisms. It is possible to present the essential theory of prophylogeny very simply and briefly (Boyden and Shelswell, 1959).

1. There has been an evolution of evolutionary mechanisms from primitive and generalized to advanced and specialized stages.

2. The original steps in the evolution of the lower Metazoa were probably accomplished by the operation of primitive genetic and reproductive processes.

3. Asexual reproduction is probably more primitive in the lower Metazoa than sexual reproduction, and is therefore more likely to have been concerned in the early stages of Metazoan evolution.

4. It is therefore incorrect to attempt to account for the remote origins of the lower animals by the operation of more recent and specialized mechanisms which probably did not at those times exist.

Now let us apply these principles to the problems of Cnidarian phylogeny, and particularly to the problem of the relative primitiveness of the polyps or the medusae in the Hydrozoa. In keeping with our theory we would compare these two kinds of adult Coelenterate in regard to their relative structural advancement and especially in regard to the kind of reproductive processes they exhibit. Those adults which reproduce predominantly by sexual means would then be considered to be more advanced than those which reproduce predominantly or solely by asexual methods. And since asexual reproduction where it is found is so generally associated with regeneration, this capacity also

TABLE 3.2

PRIMITIVE AND ADVANCED CHARACTERS OF POLYPS AND MEDUSAE IN THE HYDROZOA

Character	Polypoid Adult	Medusoid Adult
Reproduction	Mainly asexual; budding, fission, laceration, frustule formation.	Mainly sexual; occasional budding.
	Can bud polypoids or medusae.	Cannot bud polypoids.
Regenerative capacity	Exceptionally high.	Not as great.
Adult transformation	Polyp can rarely transform into medusa.	Medusa cannot transform into anything.
Structural grade of the adult	Neuromotor system relatively simple.	Neuromotor system relatively complex.

will have a bearing on the comparison of polyps and medusae. A summary of such a comparison is presented in Table 3.2.

The table brings out the fact that the Hydrozoan polyps have a preponderance of simpler and presumably more primitive characters, whether structural or reproductive. The medusa is clearly revealed in its nature as a specialized Coelenterate and therefore an unlikely candidate for the post of ancestral Hydrozoan. Such an interpretation of the life-cycle of the Hydrozoa is in agreement with the general trends in all the animal phyla which appear to have been:

1. From simple to more complex structural organization.
2. From higher to lower regenerative capacities.
3. From greater to lesser frequency of asexual reproduction.

This conclusion in regard to the primitiveness of the polyps* is obviously opposed to those views which assume that the medusa is primitive and that the short and direct life-cycle from egg to planula to actinula to medusa better represents the ancestral ontogeny. The bulk of the evidence we have presented indicates that the medusae do not now have, and probably never have had, the primitive characters which would qualify them to be the ancestors of the polyps.

Experimental evidences bearing on the relative primitiveness of polyps and medusae have recently been reported by Werner (1963). Werner points out that the marine Hydrozoa which he has reared in the laboratory for many years "show a high degree of plasticity in adjusting their development, growth, and form to such changes" as occur normally in their native environments. Thus the availability of a substratum for attachment of the stereogastrula, a normal food supply, and the temperature, all have significant effects on the life-cycles, adult form, and manner of reproduction of these Hydrozoa.

A striking example is that of the effect of temperature on the mode of reproduction of the polyps of *Coryne tubulosa*. At 14°C the colony reproduces only by the formation of new stolons and hydroids. When the temperature falls to 2°C, the hydroids bud off medusae. A return to 14°C brings the cessation of medusa formation and a resumption of stolon and hydroid formation. Further study revealed that the temperature of 6–8°C was critical and that to varying degrees abnormal

* The evidences which indicate that the Anthozoa may be the more primitive Cnidaria (Hadzi; Pantin) are to some extent pertinent here.

medusae with reduced bells and long manubria or hydroids instead of medusae were produced. Here, steps in the reduction of the medusa generation occur under conditions which are naturally occurring and they thus indicate that the corresponding evolutionary process, presumed to have occurred in several taxa of the Hydrozoa, is a likely presumption.

Generally speaking the plasticity of the hydroids appeared to be greater than that of the medusae. In one experiment it was shown that the shape of the bell of *Stomotoca ocellata*, which is at first globular, was affected by the food of the young medusa. The globular shape of the bell is retained if food is minimal; on the other hand, if food is normal a conical apical process develops. In another case the temperature determines whether the medusa of *Rathkea octopunctata* buds medusae from its manubrium or produces eggs. On the other hand, the effects of environmental conditions on the hydroids and their reproductive processes are much more striking. These experiments appear not only to support our theory of the relative primitiveness of the hydroids, but bear on the mechanisms of character determination and ontogenetic differentiation.

Cadet Hand (1959, 1963) as also Hyman (1959) agree in the view that the Hydrozoa are the more ancient class of Coelenterata, with which we also agree. However, they both select the medusa as the primitive form of the Hydrozoan stock. The facts which Hand selects in support of the relative primitiveness of the Hydrozoa among Cnidaria include (1) the greater variability of the body forms of Hydrozoa in comparison with the Scyphozoa and Anthozoa; (2) the radial symmetry of Hydrozoa; (3) the greater number of types of nematocysts in the Hydrozoa.

These facts do certainly, in our view, sustain the relative primitiveness of the Hydrozoa both among Cnidaria and in comparison with Turbellaria, but they certainly do not support the conclusion that the medusae are the more primitive adult forms for the Hydrozoa. The point (1) above, greater variability of body form, would strongly support the relative primitiveness of the hydrozooid body form rather than the medusa form.

As we have suggested, medusae have many attributes of specialization in comparison with polypes including their almost complete reliance on sexual reproduction, and their relatively poor regenerative capacities.

But before we conclude the discussion, there is one more aspect of phylogenetic method which should be mentioned, viz., the relation of metagenesis to recapitulation.

Asexual reproduction and recapitulation

The mechanisms of ontogeny and phylogeny are significantly different: in the former, the genetic code is normally constant and in the latter the genetic codes become altered. It is theoretically impossible therefore for any single ontogeny to exactly parallel a phylogenetic sequence, but an apparent parallelism in the expression of somatic characters between ontogenetic and phylogenetic sequences may occur. The genetic mechanism actually exists through which such a recapitulation would take place, if only the recent evolutionary changes are expressed at the end of the life-cycle of the descendant organisms. The fact is that recent genetic changes are not generally restricted in their expression to the terminal stages of ontogeny and therefore actual recapitulation of phylogeny by ontogeny tends to become an infrequent and uncertain phenomenon.

Now a wholly new approach to the interpretation of recapitulation is based on the inclusion in it of asexual reproduction. It is an amazing thing that apparently in all the literature of recapitulation (Boyden, 1954a; Boyden and Shelswell, 1959) no reference is made to life-cycles that involve asexual reproduction! But if our principles of prophylogeny are sound, any complete recapitulation of the early evolution of the lower Metazoa revealed in the ontogenies of existing Hydrozoa should involve life-cycles which begin in asexual reproduction. For a long time, such metagenetic life-cycles have been known but their full meaning has been unappreciated. Thus Garstang (1922) in his analysis "The Theory of Recapitulation: A Critical Restatement of the Bio-genetic Law" refers in both text and figure only to life-cycles which begin with zygotes! (Fig. 3.1). A more generalized interpretation of the relations between ontogeny and phylogeny is shown in Fig. 3.2.

Finally, let us recall the discussion of the works of Berkner and

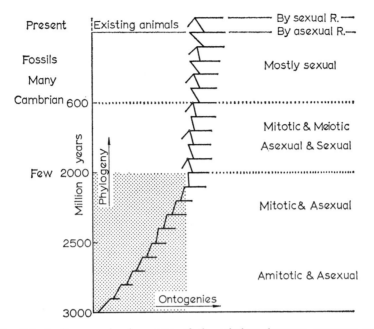

FIG. 3.1. A diagram showing some of the relations between ontogeny and phylogeny as extending back to the beginnings of evolution. The ontogenies advance to the right, successive ontogenies become phylogeny advancing upwards. Single lines of phylogeny represent successive ontogenies produced asexually. Converging lines represent phylogenetic successions resulting from sexual reproduction. The stippled area (lower left) represents stages unrecoverable in either ontogeny or phylogeny because no existing organisms retain the premitotic cell divisions characteristic of primitive evolution. Modified from Boyden (1954a). Courtesy of *Systematic Zoology*.

Marshall, referred to in Chapter 2. These authors have pointed out that benthic organisms would be protected from ultraviolet radiation whereas pelagic organisms would not, with the levels of atmospheric oxygen available in Pre-Cambrian times. By late Pre-Cambrian, sufficient protective oxygen may have been available to support pelagic jellyfish, and Glaessnar has reported such fossils. But there was nothing to prevent the origin and evolution of the hydroid forms of Coelenterata for a greater part of the Pre-Cambrian.

60   PERSPECTIVES IN ZOOLOGY

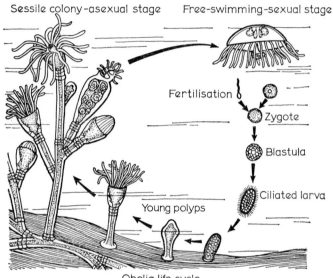

Sessile colony-asexual stage   Free-swimming-sexual stage

Fertilisation

Zygote

Blastula

Ciliated larva

Young polyps

Obelia, life-cycle

FIG. 3.2. The life-cycle of *Obelia*. Reproduced from Buchsbaum (1938), *Animals without Backbones*. University of Chicago Press. Copyright 1938 by the University of Chicago. All rights reserved. With permission. The life-cycle includes three kinds of adults, viz., hydrazooids, gonozooids and medusae, all of which may be produced by budding. Only one kind of organism is produced by sexual reproduction, viz., the original polyp from which the colony is derived. The specialized nature of sexual reproduction is revealed in its restricted reproductive capacity to a single kind of adult product.

To return to metagenesis. It is, in our interpretation, clearly evident and relatively primitive in Hydrozoa and Scyphozoa, and where it is absent as in the Trachylina it has probably been lost. Polyps better qualify as representatives of the Coelenterate ancestor than medusae for the reasons already given. In a recent report Pantin (1960) concludes that the structural organization of the Anthozoa indicates that this is the most primitive class of the Coelenterata. At the same time he finds no evidence to support Hadzi (1953, 1963) in the theory of a turbellarian derivation of the Anthozoa. On the contrary, according to Pantin, the structural organization of the Anthozoa is so essentially different from that of the Turbellaria that the former cannot be considered to have evolved from the latter. Jägersten (1959) also has recently strongly attacked Hadži's theory.

We agree with Pantin that polyps better represent the ancestral Coelenterate than medusae, but we do not agree with his interpretation of the polyps of Hydrozoa and Scyphozoa as the "immature" stages of organisms which are primarily medusae. Polyps wherever they occur are terminal stages, which may reproduce but never normally metamorphose into subsequent stages. They cannot therefore qualify as larvae, but are adults, however simple in organization they may be. Generally speaking, the chief differences among polyps in the classes of Coelenterata relate to their relative sizes and complexity and in these particulars the Anthozoan polyps seem to be more advanced than the polyps of the other classes. The advance toward bilateral symmetry in the Anthozoa has been plausibly explained by Pantin. However, the relative simplicity of Anthozoan nematocysts, compared with their greater development in the Hydrozoa seems to be the main reason why Pantin considers that the Anthozoa are the most primitive class of the Coelenterata. Whichever theory comes to have the greater evidence in its support, we assume that polyps are antecedent to medusae and that probably the Anthozoa diverged from the Hydrozoa and Scyphozoa before any medusae had been evolved. Certainly there is no evidence at present that any Anthozoan species has ever possessed medusae, much less that the whole class evolved into or from medusae and then lost all trace of their former existence. For smaller categories, however, the Trachylina may have lost polyp stages and the Hydrida may have lost medusoid stages in their evolutionary history. Finally, in regard to the phylogenetic conclusions to be drawn from the kinds of nematocysts we think a word of caution is proper. Evolution does not necessarily proceed with the same rates in all cells, tissues, and organs. The Anthozoa are the more advanced in symmetry and mesenteries, and the less advanced in nematocysts and complexity of the life-cycle. The opposite condition may be found in Hydrozoa and Scyphozoa. Comparisons of the nature of the organisms and inferences in regard to ancestry should both be as broadly based as possible.

A Hydrozoan life-cycle with metagenesis may be nicely illustrated by *Obelia*, and the corresponding cycle for Scyphozoa by *Aurelia*. Figures 3.2 and 3.3 as reproduced from Buschsbaum show the stages

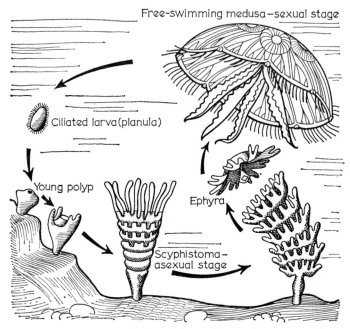

Aurelia life-cycle.(Based on various sources)

FIG. 3.3. The life-cycle of *Aurelia*. Reproduced from Buchsbaum (1938), *Animals without Backbones*. University of Chicago Press. Copyright 1938 by the University of Chicago. All rights reserved. With permission. The metagenetic life-cycle of *Aurelia* includes only two kinds of adults, viz., polyps and medusae. Budding by the polyp produces a pile of larval medusae (ephyrae) which is called the scyphistoma. Later the ephyrae are detached and mature into the gamete forming medusae.

clearly. We must emphasize several points about these cycles. The production of diverse types of adults in the *Obelia* cycle results directly from asexual reproduction—not from sexual reproduction. Thus in each case the only direct product of sexual reproduction is a polypoid organism, viz., in *Obelia* it is the original hydranth of the colony and in *Aurelia* it is the polyp which becomes the base of the scyphistoma. Budding in *Obelia* then produces additional hydranths, or hydrazooids and gonozooids. From the latter the medusae are produced again by budding. It is as though the medusae cannot "forget" their asexual heritage from "way back" though they themselves use the more advanced procedures of sexual reproduction in keeping with their more advanced morphology.

Similarly in the case of *Aurelia*, the polyp sexually produced reproduces by a process of successive budding to produce the larval ephyrae, which in turn metamorphose into the adult medusae. In the Hydrozoa the polyp stage of the life-cycle is predominant, and in the Scyphozoa the medusa stage is predominant and the life-cycle apparently more advanced.

Now the presence of comparable life-cycles in Trematoda and Cestoda (Fig. 3.2) may be interpreted as evidence of phylogenetically similar capacities, even though these instances may be cases of parallelism or convergence.* That these metagenetic cycles can be compared stage for stage is illustrated in Fig. 3.4. And in each case the interpreta-

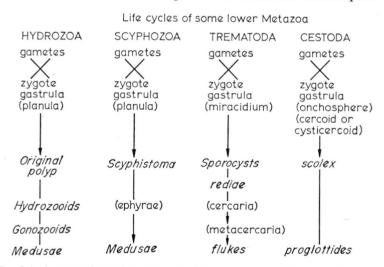

FIG. 3.4. A comparison of the life-cycles of four classes of Metazoa. The adult stages are *in italics*, the larval stages are placed in parenthesis. The connecting arrows indicate continuous ontogenies; the single lines indicate asexual reproduction; the crossed lines represent sexual reproduction. From Boyden and Shelswell (1959). With permission.

* Greenberg (1959) makes a strong case for the origin of Cnidaria and Platyhelminthes from separate groups of Protozoa.

tion is in keeping with our principles of prophylogeny. Of course the appraisal of the phylogenetic significance of the life-cycles of the parasites is made more difficult by their parasitic specializations. Nevertheless, we believe that if the interpretation is based on the definitions of larva and adult, and sexual and asexual reproduction, which we have previously used, that useful and significant conclusions can be drawn.

In the case of the Trematoda, we must be clear about the proper identification of the stages in the life-cycle. In *Fasciola hepatica*, the *egg* develops through the embryo into the larval miracidium. The miracidium metamorphoses into the sporocyst. This is a terminal and therefore *adult* stage, though of the simplest sac-like construction.* Obviously, there is a resemblance which may be significant between a simple polyp and a sporocyst. After the sporocyst come two generations of rediae, each of which is a terminal and therefore adult stage. The second generation of rediae produce the larval cercariae which ultimately metamorphose into the adult stage.

According to Hyman (1951), the digenetic life-cycle is similar to that of hydroid coelenterates, comprising a period of asexual multiplication during immature stages followed by sexual reproduction in the mature stage. It was her opinion that it is better to view such a cycle simply as a continuous ontogeny involving asexual multiplication in larval stages. I am glad to say that we agree with Hyman in noting the similarity between the life-cycles of some Hydrazoa and Trematoda, though we differ in the interpretation of these cycles. In the first place, neither the sporocysts nor rediae have the essential qualifications of *larvae* since they are terminal stages and never transform into anything. In the second place, it is incorrect to view such a life-cycle as a "continuous ontogeny", since an ontogeny is the life of a single individual and here there is a succession of individuals. And finally, since there is a regular alternation in the life-cycle of these digenetic Trematoda between a generation sexually produced, viz., that from egg to miracidium to mother sporocyst; and generations asexually produced, viz., daughter sporocysts and rediae, and finally the generation from cercariae to adult flukes, this cycle is properly referred to as *metagenesis*.

* Parasitologists incorrectly call these "larvae".

To complete the exposition of the parallelism of the life-cycles of four of the classes of the Coelenterata and Platyhelminthes requires a brief reference to the Cestoda. Here the resemblance to the meta-genetic cycle of certain Hydrozoa is less close and the interpretation therefore is more strained. However, one can see a general corre-spondence in that a generation sexually produced, running from the egg to a cysticercus and to and including the scolex, is regularly followed by generations, the proglottides, produced by budding from the scolex. This is metagenesis as the term is commonly used today.

Finally, what of recapitulation in these life-cycles? Can these onto-genies have any significant relation to their presumed phylogenies? Again, let us be sure of our terms. An ontogeny is the life-cycle of a single individual organism. A phylogeny is a succession of ontogenies in some line of evolutionary descent. De Beer's (1958) view was that phylogeny is a scale of beings arranged in a row of adult forms which descend not from adult to adult, but from the fertilized egg which develops into one adult to the zygote which produces the next. There are two matters which we find objectionable in De Beer's statement, viz., the restriction to rows of adults and the exclusion of all life-cycles beginning in asexual reproduction. In other words, the principles of prophylogeny are automatically ruled out, and hence the whole approach to the study of the evolution of the lower Metazoa by means of primitive evolutionary mechanisms is also excluded.

Now if a phylogeny is properly viewed as a succession of ontogenies in some line of ancestry, that kind of life-cycle which includes the production of different forms in successive ontogenies could provide the closest possible parallelism to the particular phylogeny involved. Such a life-cycle occurs whenever there is metagenesis. In a sense, therefore, a metagenetic life-cycle could represent a phylogeny in miniature, but differing from the real phylogeny which must have included evolutionary changes. Of course, no single ontogeny includes within its limits any evolutionary change either so this point does not deny the possibility of there being a general parallelism between the stages in a single ontogeny, or better, the stages in a metagenetic cycle, and the evolution of those organisms.

It is perfectly valid, then, to examine the metagenetic cycle for clues as to the phylogeny of the Hydrozoa or any other organisms which show metagenesis. But in a cycle where do we begin? Garstang (1922), De Beer (1958) and a host of others who have discussed this problem, begin with gametes, the most specialized cells, produced in many Hydrozoa and Scyphozoa by medusae, the most specialized adults! We believe that it is most unlikely that complicated and specialized organisms and processes antedated the simpler organisms and processes *in the lower Metazoa*. On the contrary, it seems that it would be more probable for the evolution of the Cnidaria to have followed the sequence polyps to medusae with asexual reproduction producing the increasing variety of adult types now recognized in certain of the Hydrozoa. Once the specialization had been acquired, simplifications could have followed as is generally believed to be the case where the free-swimming medusae have been reduced to simple attached gonophores. It would also be possible for the reduction of the polyp stages in the interests of a more rapid and specialized development to have culminated in the direct development of the Trachylina where, in consequence, ancestral reminiscence has been reduced to a minimum. Along the same line of interpretation we would infer that the sporocysts and rediae of the Trematoda, parasitic in invertebrates, more nearly represent the ancestors of the digenetic forms than the highly specialized flukes, parasitic in vertebrates.

For Dr. Hyman, metagenesis was a senseless idea. But it cannot be senseless to the organisms which exhibit it and have done so apparently since early Paleozoic times. Wholly aside from the light it has shed on the evolution of the lower Metazoa, it has quite obvious benefits which readily account for its persistence over hundreds of millions of years. First there is the matter of colony formation, which arises from the fact that the hydrazooid buds remain attached to the original hydrazooid. Not only are the members of the colony attached to each other, but the whole colony is attached to the substratum and can thus survive in its fixed position.

Second there is the polymorphism associated with asexual reproduction. Division of labour is an old and venerable evolutionary achievement, and the most effective way to provide diverse types of individuals for attachment, for feeding and protection, and for the two kinds of reproduction in their proper relative positions and proportions to each other has been to take advantage of the potential abilities of the budding processes to produce diverse types of adults without any change in the single genetic code available. The amounts of differentiation which occur in the whole life-cycle of *Obelia*, shown among the various types of individuals, are comparable to the onto-genetic differentiation which occurs during the development of single individuals of many other organisms, which also occur under the guidance of a single genetic code. In some respects the Hydrazoan solution to the problem of differentiation is more effective than the solution of more advanced organisms which can reproduce only sexually, for in the latter the amount of development has had to increase in the sense that ontogeny begins further back in a single cell, most often a fertilized egg, and has also to go farther to reach the greater structural complexity of the adult organisms. As Bonner (1958) has pointed out in his very stimulating book, the amount of develop-ment has had to increase in those large organisms which reproduce sexually, in comparison with those organisms which can by fission or budding processes, divide already formed organelles or tissues as in *Paramecium* or the lower Metazoa. In *Obelia*, as in *Hydra* and many other Coelenterata, the germ layers are continuous from parent to offspring as is their arrangement around the gastrovascular cavity.

Now some of the benefits of metagenesis accrue to all organisms which exhibit it, though not necessarily all the benefits to each of them. The amount of polymorphism is lesser in Scyphozoa than in Hydrozoa but the fixation to the substratum is made possible through the scyphistoma. In the Trematoda, the polymorphism is greatest and the fixation is not a problem because of the location of these internal parasites in their hosts. Finally in the Cestoda, fixation of the tapeworm in the gut is important, but the polymorphism has been reduced. There is really a great deal of evolutionary significance to the metagenetic life-cycles and to the persistence of the asexual reproduction associated therewith.

Some considerations bearing on the evolution of sex

We have covered several aspects of the theory of prophylogeny as they may bear on the evolution of the lower Metazoa, but we have not finished with those biologists who appear to believe that in the beginning there was sex. We briefly referred to this topic previously but have not finished with it. The reports of Boyden (1954a) and Boyden and Shelswell (1959) also consider it.

One of the first to make extreme claims about the primeval appearance of sex was Montgomery (1906) who made the astonishing claim that asexual reproduction was secondarily acquired in all Metazoa! That sexual reproduction should be found so readily dispensable in the lower organisms should make one doubt that it was all that necessary or advantageous to begin with. The difficulties in this view of the early evolution of sex are clearly evident in the report of Hawes (1963).

In Hawes' report it is claimed that the distribution of asexuality in the Protozoa is more "economically" explained by the hypothesis that asexual reproduction is the result of a loss of sexual reproduction which has occurred independently in different groups of Protozoa. It is also stated that the evidence speaks for this view rather than the alternative view that asexual reproduction is primitive.

What is this evidence? It is certainly not to be found in the present distribution of asexual and sexual reproduction among Protozoa. Hawes states clearly that an apparently exclusive reliance on asexual reproduction is common in the Mastigophora and Rhizopoda but on the other hand rare in the Ciliata and unknown in the Sporozoa. He goes on to say, "It is at once seen that asexuality distinguishes the simpler and sexuality the advanced orders." Again, in speaking about various orders of the Mastigophora (e.g. Euglenoidea, Dinoflagellata) he states that the well-authenticated examples of sexuality are found in extremely specialized genera, and that a similar situation is to be found in the Zoomastigina. In every case he cites, sexual reproduction is associated with specialization and not with primitiveness. Now just where is this "evidence" that sexual reproduction was primitive and has secondarily given way to asexual reproduction? The only "evidence" given relates to the deterioration of the sexual processes in some of the more specialized groups which may culminate in various types of autogamy or other modifications which interfere with genetic recombination. But let us point out that this degeneration is known *only in specialized organisms* and cannot therefore have any bearing on primitiveness! The rest of the "evidence" is mere theorizing. It is Hawes' opinion that the discontinuity of sexual reproduction cannot be assumed to be the result of a polyphyletic evolution of sex because this "commits us to the improbable belief that not only have the physiology and behaviour needed for fertilization made a number of independent but similar appearances in the past, but their entries and re-entries have been severally accompanied by the repeated evolution of the whole complicated cytological basis for sex, including meiosis. All this is so unlikely that we are bound to seek a more economical alternative."

It is our view that the actual evidence relating to the distribution of asexual and sexual reproduction in Protozoa definitely supports the primitiveness of asexual reproduction rather than the view of Hawes. We have shown that sexual reproduction and even mitosis and meiosis are much more varied phenomena than usually assumed and that the change from mitosis to meiosis is relatively small. As to the matter of recombination this too is accomplished in a variety of ways some of which are sexual and some constitute the "alternatives to sex" referred to by Haldane, Pontecorvo and others. Hawes' conclusion is definitely not supported by the actual facts he discusses—on the contrary these facts indicate the correctness of the alternative view that sex is specialized and asexual reproduction is primitive.

Dougherty (1955) believes that all surviving sexuality may be derived from the discovery of sex by some hypothetical "premoneran" ancestor. In order to avoid any wastage of words in regard to these views and my own, we need to bear in mind the meaning of the principal terms employed. My own expressed views relate to the evolution of *sexual reproduction*, which has been defined as reproduction involving gamete formation. It may be unisexual (parthenogenesis) or bisexual and in the latter case may involve self- or cross-fertilization. In our view, it is a specialized process, dependent on the existence of complex cycles of gametogenesis. We know of no actual transitions between asexual reproduction and sexual reproduction, and we agree with Darlington (1958) that sexual reproduction as it is found in organisms now is dependent on the regular occurrence of mitosis and meiosis as a preparation for fertilization.

Dougherty (1955) nowhere defines sexual reproduction but instead has defined "sexuality".* We quote:

* Surely "sexuality" is a broader term than is appropriate in this context. Normal and abnormal sexual behaviour of humans and other animals is a part of sexuality.

"I use the word *sexuality* to mean that property of organisms displayed *when they regularly establish new combinations of self-duplicating, deoxyriboenucleic acid (DNA) molecules, or aggregates thereof, either by merging of two separate cellular compartments or by the transfer of such material otherwise from one cellular compartment to another, usually of the same species*, where such molecules are part of the usual hereditary material of the cells involved, and where the transfer is not simply that transmission of hereditary material occurring when a cellular compartment duplicates itself by vegetative means; the word *sexuality* thus also comprehends *those processes, made necessary by transfer mechanisms that permit the sorting out and/or recombination of such molecules (in part, in whole, or in groups, within a cell)*."

This monstrous definition is still inadequate for Dougherty's purpose for he goes on to say that parthogenesis should be included in "sexuality". Also that he wishes to exclude any kind of symbiotic association where the symbionts have had a long independent phylogeny.

I might point out additional criticisms in that when mutations in genes occur, they would not be at first a "part of the usual hereditary material of the cells concerned" and therefore would be excluded from "sexuality". In the second place, I would be in doubt whether within homozygous lines organisms are really able to "regularly establish new combinations" of genes when all they do is to alternate between haploid and diploid numbers of the same genes. And, finally, does it help to define sexuality as "not simply that transmission of hereditary material occurring when a cellular compartment duplicates itself by vegetative means". What then are vegetative means?

Dougherty admits that his strictly monophyletic view is extreme, and he definitely fails to carry the assumed primeval sexuality from the unicellular organisms to the Metazoa. He has criticized me for failing to carry sexual reproductions from Protozoa to Metazoa but in my theory there was no such transfer. What he has called in my report a "devastating weakness" is therefore actually more applicable to his own theory since he assumes a strictly monophyletic sequence in all organisms and fails to carry through to its presumed end results. And when he states that Haldane has "effectively refuted" my theory of the early origin and significant role of asexual reproduction in animal evolution he is in error. Haldane's unsupported opinion is no effective refutation of anything! As a matter of fact, many of the processes which Dougherty would include under "sexuality" are referred to by Haldane (1950) as "alternatives to sex", being of a nature which does not permit their certain inclusion in any definite phylogenetic sequence leading to sexual reproduction in the Metazoa. We find the term "parasexual" as used by Pontecorvo (1959, p. 116) a welcome addition to the discussion. We quote from his Introduction (p. 1): "A by-product of recent research is the realization that sexual reproduction—i.e., a regular alternation of karyogamy and meiosis as shown in higher organisms— is by no means the only process for the pooling and reassorting of genetic information from different lines of descent. Though known so far only in microorganisms, novel processes of genetic recombination make it clear that some modernized version of the theory of the gene is applicable in organisms or situations in which sexual reproduction (the basis of the original theory) does not occur."*

We believe that it is useful to use the term sexual reproduction as we have done and to use the term parasexual as Pontecorvo has done. For stages prior to reproduction involving gamete formation and which are assumed to have led to sexual reproduction we may use the term presexual. Asexual reproduction is not in any way related to gamete formation so far as is known. Certainly there is no evidence even suggesting that a bud has ever been made into an egg.

Stebbins (1960, p. 201) believes that the term "sexuality" should be confined to the precise nuclear behaviour found in Eukaryota. Following Wollman, Jacob, and Hayes (1956), he would apply the term "meromixis" to all types of genetic recombination known in the Prokaryota. He believes that the extension of the meaning of well-known and generally used terms to cover newly-discovered phenomena of a basically different nature tends to confuse scientific description.

Some further comments may help to explain why we do not believe it useful to consider sexual reproduction as synonymous with proto-plasmic fusions or recombinations. In the first place there are many kinds of protoplasmic unions which have no apparent relationship to sex. These include the unions of algae and fungi in the production

---

* From Pontecorvo, G. (1959, p. 1), *Trends in Genetic Analysis*. Columbia University Press, New York. With permission.

of lichens; such cases of symbiosis as occur between algae and coelenterates and flatworms; meromixis in bacteria; mycelial fusions leading to dikaryogamy in fungi; and the union of great numbers of amebulae to constitute a large plasmodium in Mycetozoa. In the second place the occurrence of recombination is not coextensive with sexual reproduction. In autogamy and usually in diploid or polyploid parthenogenesis recombination is lacking. Even in homozygous self-fertilizing plants, it is hardly justified to speak of "recombination" occurring when all that usually happens is an alternation between haploid and diploid numbers of the same genes.

Sexual reproduction is indeed viewed much too narrowly when it is conceived in terms of fusion and recombination only. Those who have such limited views seem repeatedly astonished that sexual reproduction should continue when recombination does not occur. The fact that it so often does should suggest to them that there is more to sexual reproduction than is indicated by their "philosophy". What is this "more"?

The only universal property belonging to sexual reproduction is that when it occurs offspring are produced from unitary cells. But in this respect the same can be said to be true of asexual reproduction in unicellular organisms. The fundamental matter of distinction between sexual and asexual reproduction is that the former is a more specialized process and the latter much less so. This point of view can be clearly presented if we bear in mind the distinction between sexual reproduction which is characterized by gamete formation, and such forms of reproduction as involve meromixis or other "alternatives to sex" such as dikaryogamy resulting from hyphal fusion.

Being the more specialized process, sexual reproduction has all the advantages of specialization. Among these is efficiency. Whereas in asexual reproduction, relatively large fragments of the parental protoplasm are required for the production of daughter cells (approximately 50 per cent of the body of the parent in a *Paramecium*), very much smaller relative amounts of parental protoplasm are required in sexual reproduction. To be sure, specialization has its dangers and unspecialization may lead to survival as it has done wherever asexual reproduction has persisted in the very earliest and most primitive organisms.

The predominance of sexual reproduction in higher plants and animals and its persistence through various modifications which limit or prevent recombination is obviously not entirely a matter of producing or increasing variability but a matter of the evolution and use of specialized machinery for the production of offspring. In specialized groups of organisms, this may be the only reproductive machinery now available. In many of these, various kinds of parthenogenesis have evolved independently in the taxa concerned, maintaining the genetic codes available, which have proven successful in the business of living. Under these conditions recombination would be wasteful and inefficient. Or, in great and successful groups of plants, self-fertilization may be maintained, without apparent ill effects. All this seems quite incomprehensible to those who seem to understand sexual reproduction as equivalent to recombination and increased variability only. Obviously this is too limited a view of the matter and more perspective is necessary.

Finally, let us turn to the matter of the extreme monophyletic view of the evolution of recombination ("sexuality" in Dougherty's term). We have seen that various kinds of protoplasmic fusions occur in plants and animals. These may take place even between not closely-related organisms or among individuals of a single species. There appears to be polyphyly in such fusions, certainly there is no necessity for assuming that some "hypothetical premoneran" was the only kind of organism which could discover fusion. And even assuming that such an event took place in early evolution, how can such a process ascend through the more primitive Algae and Protozoa which show no such processes? Nor is there a shred of evidence indicating that they ever possessed such behaviour! In other words recombination phenomena are missing from the very groups of lower plants and animals which should have received and transmitted such capacities, if the views of Dougherty, Hawes and others are true. Will assumed losses account for the "missing links" in this story? Certainly not since the only evidences of "loss" relate to limitations on recombination processes which have independently evolved in several groups of *specialized* organisms only!

The occurrence of rare recombinations in bacteria has apparently been considered of decisive importance by Dougherty and Stebbins. They have ignored the possible conclusion, an obvious one, that such behaviour has been recently acquired. And after two or three billion years such evolutionary acquisitions would surely be expected. If there were great advantages to be gained from such meromixes we should also expect that it would become the standard reproductive process instead of merely occasional. Its infrequency under natural conditions surely argues against its primitive universality.

To return to our presumed early stages of the evolution of the Metazoa takes us back to premitotic and multinucleate organisms. In these, variability was great. Stabilization was acquired after mitosis was evolved. Up to this time there was no need of sexual reproduction to increase variability or adaptability, nor even to participate in the process of stabilization. It is an interesting point that several theories in regard to the origin of sex assume that the original function of sex was to make up for deficiencies resulting from mutation.* Thus Dougherty (1955) calls attention to the presumed benefits that would result from the pooling of their undamaged parts by organisms or cells which had "damaged" molecules. He seems to have forgotten that the normal undamaged molecules were available all the while to carry on the species under the operation of the forces of natural selection. A similar thought is to be found in the valuable paper of Gabriel (1960) in which it is stated that "Under conditions of high population density, cells with complementary losses might surive by cross-feeding." In fact Gabriel seems to believe that such a process was a forerunner of sexuality occurring even in premitotic organisms and leading to the establishment of regular fusions.

Similar views have been presented by Beadle and Coonradt (1944) and others. In our view it is unnecessary to seek for stabilization in early evolutionary stages through complementary "cross-feeding" especially with multinucleate organisms. They were protected by their multiple nuclei. But even in the case of haploid organisms whose life would be destroyed by lethal or undesirable mutations, the normal members of the species would be selected and would survive! For the same environmental requirements that would permit the survival of cross-feeding, which merely covers up deficient mutation, would have supported the survival of the previous normal genes! In our view, the selective value of sex came later, *after* the stabilization which resulted

* Blyth's (1835) pre-Darwinian theory of natural selection assumed that its chief purpose was to stabilize species and prevent their evolution.

from mitotic asexual reproduction. At that time, the increased variability and possible adaptability resulting from new combinations of old qualities *could* have selective value, and after mitosis the change to meiosis making sexual reproduction possible as we know it could take place. In other words, we find little potentiality in what may be termed the *sick* molecule hypothesis of the origin of sex and it has no place in our theory. Only after sexual reproduction with cross-fertilization was thoroughly established, along with the regular haploid–diploid alternation in chromosome numbers, could the protective effect of the dominant normal gene over defective alleles become significant in evolution.

In the previous chapters we have made an attempt to correct some unnatural and illogical interpretations of biological facts and to lengthen our perspective in regard to animal evolution. In the present chapter we have described and consistently applied to the problem of the origin of certain of the lower Metazoa the new principles of prophylogeny. In both chapters morphological comparisons were essential in the development of the ideas concerned and in the next chapter we propose to turn to morphology's "central conception", *homology*, for a critical analysis of its biological significance. But before we pass on let us attempt a summary in regard to primitive evolution in animals.

Summary on primitive evolution in the lower Metazoa

Studies of the adult structures and life-cycles of the lower Metazoa have thus far led to no certain conclusions and in some cases not even to a consensus of opinion in regard to their origins. It begins to be even doubtful that more studies of the same kind will be any more successful. What is needed is a new approach or several such new approaches to these old problems. One of these has been introduced and discussed as "Prophylogeny", which may add some perspective and greater understanding to the study of primitive origins. Another new approach, and possibly the ultimate one, is through the avenues of molecular biology, wherein comparative biochemistry, physiology and serology will make their contributions to the study of the natures of organisms as revealed by their most intimate structures and functions. We shall discuss these newer approaches in later chapters, for they all have a bearing on Systematics whether or not they are believed to be revelations of phylogeny.

The essence of our new look at the origin of the Metazoa can be stated in terms of a simple principle, viz., that *the primitive representatives of the older phyla evolved by primitive mechanisms.* The first steps in prophylogeny, therefore, require that we study the lower Metazoa, examine their genetic and reproductive processes and select those which are simpler and presumably more primitive. When this was done it was seen that both the distribution and nature of asexual reproduction indicate that it was the sought for more primitive type of reproduction both in Protozoa and Metazoa.

This conclusion provided a new basis for the comparison of the life-cycles of Cnidaria and Platyhelminthes, which basis had been previously entirely neglected; for all previous students of the lower Metazoa had belittled or completely ignored the phylogenetic implications of asexual reproduction. On the contrary, for *primitive* evolution, it is the specialized sexual mechanisms which should be reduced in importance.

Thus it is asexual reproduction which even now plays the greater role in the life-cycles of the metagenetic Cnidaria and Platyhelminthes and these metagenetic cycles appear to throw new light on the evolution of these two phyla. When asexual reproduction is given just consideration in these phyla there emerges a theory of the evolution of the lower Metazoa in which the following conclusions mutually support each other and fail to support the theory of Hadži.

1. Cnidaria are more primitive than Platyhelminthes.

2. Within the Cnidaria polyps are more primitive than medusae.

3. Many Cnidaria and Platyhelminthes have comparable metagenetic stages in their life-cycles.

4. Within the Platyhelminthes (Trematoda), the sporocysts and rediae more nearly represent the ancestral types than do the adult flukes.

Evidences to support the above theory are to be found in many facts of structure, life-cycles and behaviour of the organisms concerned and these evidences may no longer be safely ignored.

CHAPTER 4

MORPHOLOGICAL COMPARISONS: HOMOLOGY AND ANALOGY

Introductory

THE PRECEDING chapters have touched upon the results of morphological comparisons as in the matters of the homology of the cells of Protozoa with those of Metazoa, and in the comparisons of polyps and medusae. It was possible to present the results of these comparisons in a relatively simple and direct way in the interests of economy of words and concentration on the matters of immediate importance. But morphological comparisons are, generally speaking, far from simple and the implications of such comparisons are still the subject of much discussion. We must now therefore critically re-examine the history, meaning, and implications of "morphology's central conception" (Huxley in De Beer, 1928). In subsequent chapters we will treat the subject of biochemical and especially serological comparisons equally critically.

Historical review of the concepts and terms

The concept of a morphological correspondence, part for part, between the bodies of two organisms is older than the term homology. One of the earliest expressions of this concept is that of Belon (1555) who figured the skeletons of a bird and of a man indicating what he considered to be the corresponding parts by similar designations (Fig. 4.1.) Belon succeeded best in his interpretation of morphological correspondence where the relative position of the parts is easiest to establish, as in the limb bones. Considerably later P. Camper, Dutch naturalist and comparative anatomist, in 1778 gave lectures in Paris "on the analogy which exists between the structure of the human

77

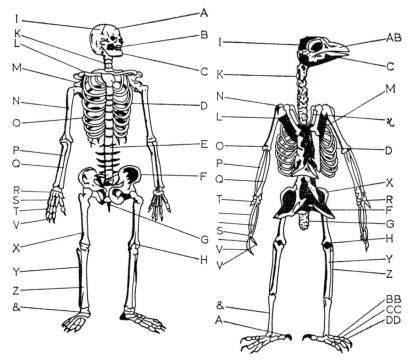

FIG. 4.1. The skeleton of a man and a bird compared by Belon (1555). The corresponding parts are identified by the same letters, the correspondence being suggested mainly by relative position and connections. The signification of the letters and symbols is given in the original French and in English translation by Hall (1951). Belon apparently left three of the lead lines without designation in the bird skeleton's wing. Copied, with permission, from Singer, C. (1959), *A History of Biology*, 3rd ed. Abelard-Schuman, London and New York.

body and that of quadrupeds birds and fish". According to Spemann (1915) Goethe was there and saw that Camper illustrated on the board how different vertebrates could be converted into one another by the modification of corresponding parts without change of relative position and connections of these parts. In his lecture, Camper made fun of the concept of angels with both arms and wings—the principles of correspondence in all vertebrates indicating that you might have arms or wings in their proper position, but not both.

Goethe in 1784 made good use of the principle of constancy of relative position among corresponding parts in his discovery of the intermaxillary bone in man. Though there appears to be no intermaxillary bone in the adult human skull, the upper incisors which always appear on the intermaxillary of most mammals are there. How could the bone be lost if the teeth are still there? Goethe thought it must be there and later (1822) found that in immature skulls it was. Actually, though Goethe was ignorant of this, Fallopio had come to a similar conclusion in 1561 and Vicq d'Azyr in 1780.

Early in the nineteenth century Geoffrey St. Hilaire (1818) developed what he called his "Theory of the Analogues" which stated that nature works always with similar materials. The beings of the same group are composed of corresponding ("analogous") parts which occupy the same relative positions in the bodies of these organisms. "An organ is more often altered, atrophied or annihilated than transposed."* It is obvious that here the word analogy has the same meaning as homology as used later by Owen (1843). In fact the human capacity to confuse with words is all too fully documented by the history of the uses of the terms homology and analogy and of the interrelations between them. That morphology's "central conception" should be treated in this way is a clear demonstration of the appropriateness of of the criticism by Woodger (1929); that nothing is more striking in biology than the contrast between the brilliant skill, ingenuity and care bestowed upon observation and experiment, and the carelessness in regard to the definition and use of concepts in terms of which its results are expressed. In fact we find here many further instances of what was referred to as "un-natural history" in the first chapter.

According to Owen (1847, 1848, 1866) the term "homologue" had been in use prior to 1843 but "vaguely or wrongly". Huxley (1894) stated that the term was well known and was in use even in its "present" sense as far back as 1800, but Lankester (1899) praised Owen for giving "precision and currency" to the term homology and stated, "It is not easy to exaggerate the service rendered by Owen to the study of Zoology by the introduction of this apparently small piece of verbal mechanism; it takes place with the classificatory terms of Linnaeus". Owen took no credit for the actual "introduction" of the term homology but the

* A literal translation.

record is clear that he deserves praise for having brought it into general usage. Owen's first definitions of analogue and homologue are to be found in the glossary of his *Lectures on the Comparative Anatomy and Physiology of the Invertebrate Animals* (1843):

Analogue. "A part or organ in one animal which has the same function as another part or organ in a different animal."

Homologue. "The same organ in different animals under every variety of form and function."

Owen's 1843 definition of homologue is of course incomplete—it fails to specify the criteria by which the relation of homology is to be recognized. These criteria were soon supplied for the relation of homology. "These relationships are mainly, if not wholly, determined by the relative positions and connections of the parts, and may exist independently of form, proportion, substance, function, and similarity of development" (1848). In a later statement (1866) Owen says, " 'Homological Anatomy' seeks in the character of an organ and part those, chiefly of relative position and connections . . ." thus indicating that his more mature view included the realization that similarity in relative position and connections could not serve as the sole basis for the determination of homology. Owen continues, "This aim of anatomy concerns itself little, if at all, with function." Owen's term "function" referred to the use of the part rather than any intrinsic physiological mechanism, as is evident from the following quotation which also makes clear the distinction between homology and analogy:

"But homologous parts may be, and often are, also analogous parts in a fuller sense, viz., as performing the same function; thus the fin or pectoral limb of a porpoise is homologous with that of a fish, inasmuch as it is composed of the same or answerable parts; and they are the analogues of each other, inasmuch as they have the same relation of subserviency to swimming" (1848).

It is clear that Owen was aware of two major kinds of correspondence between the parts of the bodies of organisms: (1) in structure, determined chiefly by relative position and connections, and (2) in use to the organisms. These kinds of correspondence are largely independent of each other and thus sets of organs may be:

1. homologous and analogous
2. homologous and non-analogous

3. non-homologous and analogous

4. non-homologous and non-analogous.

The use of the terms as Owen defined them thus leads directly to the alternatives expressed above (Boyden, 1943a) and we doubt that it is possible to improve on their clarity. But in biology the comparison of structure becomes difficult because of the great variety in the organization of parts and the end-point of the relation of homology is therefore difficult to establish. Simpson (1959) has repeatedly maintained that what constitutes an essential structural agreement in Owen's sense can only be intuitive or subjective. However, the problem of setting limits to the relation of homology has been simplified to a certain extent by the redefinitions of homology which limit it to correspondences within major morphological types (Jacobshagen, 1925; Naef, 1926; Kälin, 1946) but the problem of determining how many and what types there are still remains. We shall have need to return to the development of the meanings of homology again but let us look more particularly to the meanings of analogy which have been even more confused (Boyden, 1947; Haas and Simpson, 1946).

It seems that Aristotle (*Historia Animalium*) used analogy as Owen did so much later, to signify a similarity in the use of parts. But the nineteenth-century French zoologists used analogue in the same way that Owen used homologue (Saint-Hilaire, Cuvier, Dumeril). MacLeay (1825) used *analogy* to mean a superficial resemblance as contrasted with the fundamental resemblance of *affinity*. (But without any necessary implication of common ancestry in the relationship of affinity.) The crowning bits of confusion came later after Strickland (1846) and Darwin (1859) for then analogy was often used to mean a superficial similarity in structure, together with, or "*due* to" a similarity in use. That this chimaerical usage of analogy has been confusing for a long time is nicely illustrated by the following quotation from Turner (1899, p. 830).

"When two or more parts or organs correspond with each other in structure, relative position, and mode of origin, we say that they are homologous parts, or *homologues*; whilst parts which have the same function, but do not correspond in structure, relative position, and mode of origin, are analogous parts, or *analogues*. Homologous parts have therefore a morphological identity with each other whilst analogous parts have a physiological agreement. The same parts may be both homologous and analogous as the forelimbs of a bat and a bird, both of which with the same fundamental type of structure, are subservient to flight."

There is surely something wrong with the meaning of terms when zoologists so readily confuse and contradict themselves in the space of a single paragraph! Let us summarize the various meanings of homology and analogy and then proceed to a discussion of the mechanism of homology and its biological implications.

Some meanings of homology

1. An essential structural correspondence. (Owen, 1847, 1848.)
2. An essential structural similarity due to common ancestry. (Darwin, 1859; Haeckel, 1866; Gegenbaur, 1878; zoologists generally.)
3. The reaction occurring between any antigen and the antiserum produced in response to it. (Kraus, 1897, and serologists generally.)
4. The relationship between chromosomes which pair during meiosis or between genes which act as alleles (geneticists and cytologists).
5. Series of parallel mutations or varieties in related species. (Vavilov, 1922.)
6. The relationship among organs composed of the same fundamental constituent parts incorporated in a single body plan regardless of form and functional differences. (Jacobshagen, 1925; Naef, 1926, 1927.)
7. Any structural similarity due to common ancestry (De Beer, 1928; Shull, 1929; Haas and Simpson, 1946).
8. Any similarity, whether structural or functional, which is of common evolutionary derivation. (Hubbs, 1944.)
9. Any tissues in an organism stimulated by the sex hormones characteristic of that sex, and inhibited by the sex hormones of the opposite sex. (Moore, 1944.)

Some meanings of analogy

1. An essential structural similarity. (Geoffrey Saint-Hilaire, 1818.)
2. A superficial structural similarity. (MacLeay, 1825; Strickland, 1846; Jacobshagen, 1925; Nowikoff, 1935.)

3. A functional similarity especially in the use of parts with or without essential structural correspondence. (Owen, 1843, 1847.)

4. A superficial structural similarity due to use. (Darwin, 1859; many modern texts.)

5. Parallel variations in related species including *homologous* characters. (Darwin, 1859.)

6. Functional or physiological agreements. (Haeckel, 1866; Lankester, 1870; Gegenbaur, 1878.)

7. A superficial similarity, structural or functional, not due to common ancestry. (Hubbs, 1944.)

8. A similarity in use of structures or functions which are of independent origin and history. (Hubbs, 1944.)

9. "The twin concepts of homology and analogy have nothing to do with similarity of features; they are only associated with common origin versus non-common origin." (Bock, 1963, p. 269.*)

10. "Homologous features (or states of the features) in two or more organisms are those that can be traced back to the same feature (or state) in the common ancestor of these organisms." (Mayr, 1969.†)

11. "Analogous features (or states of the features) in two or more organisms are those that are similar but cannot be traced back to the same feature (or state) in the common ancestor of these organisms." (Mayr, 1969.‡)

It is our opinion that there should be some effort made to "conserve" terms and their definitions, especially when a term is still needed to convey its original meaning. This conception of a priority of meanings could help to prevent the misapplication of a term, good in its original meaning, for wholly unrelated kinds of correspondence. If such a practice had been adopted we would certainly have a more consistent use of words and a more decent respect for their history, and we would hope to avoid such flagrant violations of meaning as, for example, implied in Moore's usage. But the practical question is what usages shall we now recommend? The answer to this question should depend

* From Bock, W. J. (1963), *American Naturalist* **97**, 265. With permission.
† From Mayr, E. (1969), p. 85, *Principles of Systematic Zoology.* Copyright McGraw-Hill, Inc., New York. With permission.
‡ From Mayr, E., *op. cit.*

on our understanding of the mechanism of homology and the criteria which may properly be used for the demonstration of the relationship thereby implied. Following Darwin there was a marked change in the definitions and it was then that the habit of looking backward for evidences indicating the relationship of homology took hold. Now of course a Palaeontologist must look backward in his studies but is it absolutely necessary that a Neontologist should do so? And what about the Systematist? Must he approach the study of existing organisms "from way-back"? This is not an idle question, and though, as Simpson points out (1959), there is now a consensus of opinion which supports the definition of homology as "resemblance based on ancestry" there are a few who still maintain that it is unwise to define a relationship in terms of an inference from that relationship (Woodger, 1929, 1945) and omit all reference to the actual criteria by means of which the inference is to be justified (Boyden, 1947, 1969). Furthermore, it may be doubted that Darwin did provide a "theoretically objective criterion of homology" as Simpson (1959) claims, for even if the common ancestry did exist the supporting evidences would be drawn from an appraisal of the amounts and kinds of structural correspondence among the structures compared, together with the space–time distribution of *similar* fossils if such are available. But the mechanism of homology should first be understood in so far as current information permits and then we shall reconsider the more useful ways in which to define homology.

The mechanisms of homology in brief

We come here face-to-face with what may be called the greatest biological problem viz., "what determines the characteristics of organisms?" When we can give a correct and full answer to this question for particular characters we may then be able to describe the mechanisms of homology for such of those characters as are (1) corresponding, that is essentially similar in the criteria referred to by Owen, or (2) "due to common ancestry" as required by the consensus of current opinion defended by Simpson. Since it is actually impossible to meet the second requirement except by passing through the first, we shall now consider the mechanisms which determine the characters of organisms. Later we shall need to return to the criterion "due to common ancestry" when the implications of homology are critically analyzed.

The mechanism of homology is the mechanism of heredity. Many decades of intensive study have shown that genes and cytoplasm interact to determine essential similarities of all kinds, structural, functional, and behaviouristic. In fact we may justly conclude that the specific nature of organisms is due primarily to inheritance. This does not mean that environment is óf no effect in the modification of the expression of hereditary characters, but such modifications generally do not alter the "specific nature" of the organism in the systematic sense.

The physical basis of heredity, as of life, is organized protoplasm of two major kinds, viz., nucleoplasm and cytoplasm. By both asexual and sexual reproduction, this protoplasmic organization is transmitted from the parental generation to the offspring. In fact, the only way in which one can define heredity and cover all its varied manifestations is to say that it is "the transmission of organized protoplasm from parent to offspring with its innate capacities" (Boyden, 1947). Further, it has now become common knowledge that the genes and the cytoplasm interact with each other and with the environment to bring about the marvellously integrated sequence of events leading to the adult character expressions. In more detail it was generally held until recently to be true that: (1) each gene generally has manifold effects on the body, (2) that each complex organ is conditioned by the interaction of many different genes, (3) that during development the genes usually do not differentiate but the cytosome does, (4) that the mitotic mechanism generally distributes the same numbers and kinds of genes to all cells of the body. As far as the first two statements are concerned the evidences available now would seem to support them. This would still be true even if the one-gene–one-enzyme hypothesis (Beadle, 1946) were correct, for the alteration of the action of one enzyme in an early developmental stage could still have manifold effects on the expression of adult structural characters. But a revision of statements (3) and (4) above may be indicated because the experimental studies of King and Briggs (1955) and Briggs and King (1957) appear to show changes in

D

the nuclei of developing frog embryos do occur. In brief, nuclei of embryonic cells of *Rana pipiens* of various early stages were transplanted into enucleated eggs. Control experiments in which nuclei from undifferentiated early gastrula cells were used gave good results leading to the development of apparently normal complete blastulae in 41 per cent of the cases and even to the production of 10–12 mm larvae in 35 per cent of the cases. When the nuclei were taken from late gastrula chorda-mesoderm cells or from the floor of the archenteron (presumptive mid-gut) a much higher proportion of abnormal embryos was the result and a smaller proportion reached the 10–12 mm larva stage. These results suggested that there was a progressive specialization of nuclear function during cell differentiation (King and Briggs, 1955).

In a further study (Briggs and King, 1957) nuclei of endoderm cells taken from the floor of the anterior midgut at various stages were transplanted into enucleated egg. When the nuclei had been obtained from the late gastrula the majority of the recipient eggs (63 per cent) cleaved and developed into partial or complete blastulae. Though the complete blastulae (50 per cent of the recipient eggs) were normal in appearance they usually showed marked abnormalities in later development. The arrested embryos seemed to be normal in the inductor system (notochord and somites) and in endoderm tissue, but showed marked deficiencies of ectodermal derivatives. The nuclei obtained from these endoderm cells showed progressively greater abnormailty, the older the embryo from which they were obtained. Thus when the endoderm nuclei were obtained finally from the tail-bud stage, only 17 per cent of recipients, eggs were able to cleave, and most of those that did, showed abnormalities in the blastula or early gastrula stages. A control experiment showed that the introduction of cytoplasm of embryonic stages alone into normally developing eggs produced no abnormalities. These results appeared to justify the conclusion of Briggs and King that endoderm nuclei undergo stabilized or irreversible changes during differentiation, which become progressively greater with increasing development of the embryo. One possible cause of a systematic error in the interpretation of these results might be that the nuclei obtained from the later embryonic stages were subjected to a progressively more destructive or prolonged treatment with the reagents used in preparing the cells (trypsin and versene) from which the nuclei were obtained. The data reported do not dispose of this possibility however unlikely it might be.

Several subsequent studies by Gurdon (1962, 1963, 1964, 1967, 1969), Gurdon and Uehlinger (1966), Laskey and Gurdon (1970) and others have clearly shown, however, that some of the nuclei from even fully differentiated tissues and organs have retained their original genetic potential for the determination of all kinds of normal cells. The following quotation from Gurdon (1967) nicely states the situation:

"The extraordinary capacity of plants to achieve complete regeneration from single somatic cells (Braun 1959; Sinnott 1960) demonstrates that essential genes are not lost in the course of normal plant cell differentiation. In animal development, however, there are clear cytological examples of the loss of chromosomes both by elimination, especially in certain insect families (described in White, 1954), and by diminution followed by elimination in *Ascaris* (Boveri, 1899). Probably the most direct evidence concerning genetic losses in differentiating animal cells is provided by nuclear transplantation experiments in Amphibia. In all such experiments it is found that nuclei transplanted from differentiating cells support normal development less often than do nuclei from embryonic cells. There are many possible explanations for this effect (Gurdon, 1963), and the experiments have not yet shown that genes are actually lost in the course of normal cell differentiation. On the contrary, nuclear transfer experiments have given definite evidence that in many cases the nuclei of differentiating cells can support normal development and therefore have not lost any essential genes. . . . Gurdon and Uehlinger (1966) have shown that differentiated cells of *Xenopus* sometimes contain nuclei genetically equivalent to germ-cells, since a few nuclei of tadpole intestine cells transplanted into enucleated eggs have permitted development of these eggs into fertile male and female frogs. It seems evident that essential genes are not lost in the course of normal cell differentiation, though the possibility has not yet been excluded that a cell may contain many copies of certain genes and that the number of these might be reduced during development."*

* From Gurdon, J. B., "Control of gene activity during the early development of *Xenopus laevis*". In Brink, R. A. (editor), *Heritage from Mendel*. Copyright 1967, University of Wisconsin Press, Madison, Wis. With permission.

Evidence of functional changes in genes during ontogeny is to be found in the work of Breuer and Pavan (1955).

These authors studied the behaviour of the polytene chromosomes of *Rhynchosciara angelae* during larval stages. With this exceptionally favourable material it was possible to show that certain bands in the salivary chromosomes and in the Malpighian tubules develop enormous bulbs at particular larval stages. These bulbs were associated with the production of large amounts of DNA. At later stages the bands would return to normal appearance. These observations suggest changes in functional activity of the genes located in these bands during larval development.

These experimental findings do not as first thought require a revision in the description of the mechanism of homology as given before (Boyden, 1935, 1943, 1947). In these earlier accounts the assumption was made that during ontogeny the genes were normally constant and the nuclei were undifferentiated (except when tissue cells became polyploid or diminished). The evidences from experimental redistribution of nuclei in the early stages of amphibian ontogeny and from regeneration studies indicated that such was the case. Now the newer experimental data do not deny the validity of the old data, for certainly where a part of an organism can regenerate the whole organism this can only mean that the whole genetic code was available in the cells which were capable of this regeneration. Therefore there was either no differentiation of genes during the development of the various parts of the regenerating organism, or such differentiations as may have occurred were reversible. But it does not even require experimental studies in regeneration to establish this, for wherever asexual reproduction can occur naturally, the same conclusion is warranted! Thus in all those organisms previously listed as having the capacity for asexual reproduction, viz., the Protozoa and many of the lower Metazoa and even the Tunicata, the earlier statements regarding the mechanism of homology are still valid. In other words it is only in the more specialized phyla which have lost the capacities for asexual reproduction, and much of their potentiality for regeneration as well, that an irreversible differentiation of nuclei during ontogeny *could* take place in somatic cells. The primitive mechanism of homology was, therefore, based on the interaction of constant genes with the organized and differentiated cytoplasm during the ontogeny of each organism. But just as the whole genetic code had to be retained in all cells from which a new individual could be produced by budding or fission, so enough of the organism's fundamental cytoplasmic organization had to be likewise retained, or reacquired by the reversal of any changes which had occurred during that ontogeny.

The mechanism of homology in organisms which can reproduce asexually

The *primitive* mechanism of homology can therefore be illustrated in terms of the ways in which genes and cytoplasm interact for the two kinds of homology, serial and special (Boyden, 1947).

The mechanism of resemblance in homologues
1. Serial homologues—the same genes interacting with the similar parts of the cytoplasm within the limits of one egg or asexual product, leading to the production of one embryo or one adult organism.
2. Special homologues—the same or similar genes interacting with the same or similar parts of the cytoplasms of more than one egg or asexual product, embryo or adult organism.

The mechanism of difference in homologues
1. Serial homologues—the same genes interacting with the somewhat different parts of the cytoplasm of one egg, asexual product, embryo, or adult.
2. Special homologues—different genes interacting with the same or different cytoplasms of more than one egg, asexual product, embryo, or adult.

Now if this is a valid general statement of the apparently primitive mechanism of homology as it operates in the Protozoa and the lower Metazoa, how does it need to be altered to describe properly the more specialized mechanism of those few organisms where nuclear changes do take place in ontogeny? The revised general statement would have to be so framed as to admit of nuclear changes during the ontogenetic differentiation of somatic cells such as occur in: (1) Ascarids following chromatin diminution, or (2) Diptera or other insects during the development of salivary gland and other polytenic cells, during the loss of male chromosomes or parts of chromosomes, especially of X chromosomes of paternal origin. Sonneborn (1967) has reviewed many such cases of unusual chromosome behaviour occurring in Insecta or other Arthropoda. In these organisms there may be some regenerative capacity for appendages but no such capacity to regenerate a whole organism from a fragment. Therefore it is not essential that the whole soma retains whole sets of chromosomes—but only those chromosomes or parts of chromosomes concerned in the determination of essential somatic characters. On the other hand, the germ line must retain the whole of the necessary genes and cytoplasms, of course to the extent that genes are replicated many times; as Britten and Kohne (1968) and Britten (1969) have made clear, some of the redundancy could presumably be lost without any subsequent loss of characteristics.

Speaking of cytoplasms, Sonneborn (1967) has beautifully shown that certain of the surface structures of *Paramecium aurelia* and tooth characters of the shell of *Difflugia* or *Euglypha* have an autonomy of their own and can direct their own reproduction, can mutate, and perpetuate the mutations, independently of nuclear determinations. This knowledge also has a bearing on mechanisms of homology as it eliminates the nuclei from any specific influence on these characters. But these are relatively rare instances in the whole domain of homology mechanisms. Let us pass on then to sketch the mechanisms responsible for homologies in organisms which reproduce only sexually.

The mechanism of homology in organisms which reproduce only by sexual reproduction

The mechanism of resemblance in homologues

1. Serial homologues—the same or similar maturation stages of the same genes interacting with similar cytoplasmic materials within the limits of one egg, embryo, or adult.

2. Special homologues—the same or similar mutations of the same genes interacting with similar cytoplasmic materials of different individuals.

The mechanism of differences in homologues

1. Serial homologues—similar or different maturation stages of similar genes interacting with somewhat different cytoplasmic materials within the limits of one egg, embryo, or adult.

2. Special homologues—different genes or different maturation products of the same genes interacting with somewhat different cytoplasmic materials of different individuals.

Further description of the mechanisms of homology

Our description of the primitive and advanced mechanisms of homology was presented in general terms only and is of course very incomplete. The complete description of this mechanism would require knowledge of the details of all the interactions of genes and cytoplasms during ontogeny. This knowledge is increasing due to the development of several instruments capable of revealing much more about the fine structure of protoplasm. The electron-microscope has been particularly helpful and we may hope for gradually increasing insight in consequence. To be sure we are still far from adequate understanding in any case. In reference to the relatively simple organization of bacterial cells ("protocells") Picken (1960) says, "From these naive considerations can be seen at least, how remote is the relation between the molecular structure of the genetic system, and the characteristics of the bacterium known to the microscopist; of the bacterium it is no exaggeration to say that all its morphological attributes are emergent in the sense that none is directly related—so far as we can see—to the linear organization of the genome."* Slow as added steps in understanding the details of the mechanisms of character expression, and thus of homology, may be, we feel that definite progress is being made in experiments which are capable of revealing: (1) the time and place at which particular proteins or antigens appear in ontogeny, (2) the detailed organization of both nucleus and cytoplasm, (3) the nature of the interdependence of nuclei and cytosomes.

* Developments of the last decade have provided greater knowledge, at least as far as protein structure is concerned.

The roles of genes and of cytoplasms in the heredity of homologous characters

As to the interdependence of nuclei and cytosomes it has long been held that the nucleus was the primary source of determination, and the cytoplasm was secondary. Thus Muller (1955, 1967) holds the extreme view that a single gene antedated all other expressions of the living substance, and many geneticists have appeared to assume that Mendelizing characters are determined to a decisive degree by genes, and that the cytoplasm was generally of much less decisive influence. This conclusion seems at first sight to be correct, since the substitution of alleles may lead to marked changes in the structures and functions of hybrid organisms. Because so often reciprocal crosses give similar results (except of course where sex-determining mechanisms come into play) it was held that the cytoplasm was not concerned in the determination of these alternative characters. However a very different interpretation is not only possible but probable. In all such cases the individuals are closely related in spite of their gene differences, and if the cytoplasms are more conservative than the genes (Boyden, 1943a) the equality of reciprocal crosses was to be expected and at the same time would not at all justify the conclusion that the cytoplasm was insignificant in the expression of Mendelizing characters.

It has long been known that the most conservative criterion of structural homology is relative position and connections. Thus Geoffroy Saint-Hilaire wrote (1818, p. xxx), " . . . un organe est plûtot altéré, atrophié, anéanti, que transposé." And Owen (1848), as indicated in his writings, considered that the criteria of homology were those " . . . mainly, if not wholly, determined by the relative position and connection of the parts . . .". In his later statement (1866) he spoke of homologous characters as being recognized "chiefly by relative position and connections", which is a truer statement since relative position alone, in organisms of different major construction, would not and should not lead to a conclusion of homological correspondence. Now this prepares us for the task of analyzing the hereditary mechanism for evidences bearing on the relative roles of genes and cytoplasm in the determination of: (1) the structures of the homologous parts and (2) the relative position of those parts.

It may well be that the mechanisms of these two kinds of correspondence are significantly different, as was suggested in a previous report (Boyden, 1943a). The evidences that gene substitutions can bring about relatively enormous structural changes in organs are overwhelming. It has been derived from many kinds of organisms, plant and animal. Furthermore these organisms in which gene substitutions can occur must be closely related and to some degree interfertile since the classical methods of genetic analysis require the production of at least a first hybrid generation. Even among such organisms the substitution of alleles can alter, slightly or markedly, the organs affected and may even lead to their complete or practically complete removal. Thus, multiple allelic genes at the vestigial locus in *Drosophila melanogaster* can reduce the wings by many degrees or lead to the mere stump of a wing characteristic of the "no wing" expression. Genes at other loci produce other, sometimes equally drastic, modifications. Thus the eyes are subject to similar genetic alterations. But in none of these cases have the organs been changed in regard to their relative position in the body! A thousand genes may thus alter organs *in situ* to one which may produce an apparent change of relative position! It lies close to hand therefore to tentatively conclude that relative position in the body of an organism is not the primary business of the genes and must therefore be the primary business of some other part of the genetic mechanism, viz., the cytoplasm (Michaelis, 1954).

Evidence to support this conclusion has come from an unexpected source. In his studies of nuclear transfer in Amoeba, Danielli and his associates (1955) and Danielli (1959) have discovered that the nuclei and the cytoplasms do interact and affect each other in significant ways, and that the cytoplasm is capable of character determination, as well the nucleus. Thus, after transferring the nuclei from *Amoeba proteus* immediately to the enucleated cytoplasm of *Amoeba discoides* it was found that the cytoplasmic influence is always as strong as the nuclear influence, and for most of the characters studied it was actually the stronger. However, the studies with toxic antisera showed that the nuclear influence was the greater on the formation of the antigens concerned.

These results lead to a working hypothesis in regard to the roles of nucleus and cytoplasm in character determination. In essence it appeared that the nucleus determined the specific character of the macromolecules which are formed in a differentiated cell, whereas the organization of the macromolecules into functional and structural units was determined mainly by cytoplasmic inheritance.

In the later study, Danielli (1959) has further facts and suggestions which appear to have an important bearing on the mechanism of homology. He points out that, even though in their studies of the roles of nuclei and cytoplasm in nuclear transfer experiments, the cytoplasms have the predominant effect, there is no general reorganization of the cytoplasm in the binary fissions which follow. On the other hand when fertilization occurs there may be and appears to be a profound reorganization. Thus it is possible that at each fusion of gametes or other form of mating, cytoplasmic determinants are formed anew as a result of nuclear activity and under nuclear determination. If this is actually the case we have then a further consequence of the difference between the mechanisms of homology for asexual reproduction, the presumed primitive one, and for sexual reproduction, the derivative and specialized one. And since in sexual reproduction the entire genetic code must be incorporated into a single highly specialized cell, the ovum, and a whole gene code is usually transmissible through an even more specialized cell, the spermatozoon, it is probably to be expected that the cytoplasmic organization must be reduced to its most condensed and elementary, yet *specific* form.

Now what is the nature of this cytoplasmic organization as revealed by modern studies?

We have indicated that heredity may be best defined as "the transmission from parent to offspring of organized protoplasm with all its innate capacities". The nuclear organization, at least the linear order of the genes, is well established. But in what aspects of cytoplasmic organization resides the capacity to conserve the relative positions and connections of homologous organs? And how are the structural expressions of the cytoplasm determined by its interaction with the genes? The whole of embryology and modern cytology may be concerned in providing an answer to these simple but fundamental questions. "There is no denying that development remains one of the largest collections of unsolved problems in biology today" (Bonner, 1958,

p. 1*). We have not as yet explained how bilateral symmetry is determined!

Cytoplasmic organization in asexual reproduction

The transmission of organized protoplasm from parent to offspring is a relatively obvious occurrence when the offspring is a fission product derived from the parent. It is in fact more to be expected that the products of such reproductive processes should resemble each other and the parental organism than that they should show differences. In these cases, already-formed structures may be passed on to the offspring as in Protozoa, or major features of organization much as the germ layers are continuous between parent and offspring as in budding Cnidaria. The details of such processes as occur in the binary fission of Paramecium and other Ciliates have been discussed by Bonner (1958), Lwoff (1950) and others. Here, some of the structural differences in the individual are related directly to the organization of the cytoplasm rather than to the nucleus. In fact it is pointed out by Bonner (*op. cit.*, p. 49) "that the cytoplasm of ciliates had reached such a state of complexity that in a sense it had lost some of its power of development ...". Further, in commenting on the extent of the surface polarity and other aspects of the organization of the cytosomes, Bonner states that it is quite understandable how it is more efficient to pass on this organization than to recreate it for each generation. Now if "*the function of development is the establishing of differences in different parts*" (Bonner, *op. cit.*, p. 39), then it is true that such organisms have apparently lost some of their powers of development.†

Sonneborn (1963a) has recently discussed the current theory of nucleo-cytoplasmic interactions involved in character determination and has called attention to the need of its extension to include parts of the cytoplasmic organization. This need has become evident in the study of variations in the structures of cortical areas in *P. aurelia*. These cortical areas perpetuate their own structural organization, independently of any differences in genes, or even in the endoplasms

---

* From Bonner, J. T. (1958), *The Evolution of Development*. Cambridge University Press. With permission.

† From Bonner, *op. cit.*

beneath them. For this ordering role of cytoplasmic structure Sonneborn proposes the term "cytotaxis", which he considers an essential part
of the machinery by which a cell can make another cell. And speaking

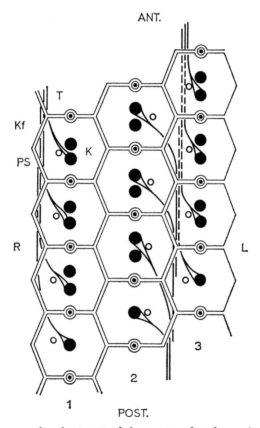

Fɪɢ. 4.2. Diagram of a short part of three rows of surface units on a cell of
Paramecium aurelia, possessing one inverted row (2) between two
normally oriented rows (1 and 3). The symbols ANT., POST.,
R., and L. indicate the cell's anterior, posterior, right and left respectively. T, trichocyst; K, kinetosome; Kf, kinetosomal fibre; PS, para-
somal sac. Reproduced from Sonneborn, T. M., Fig. 19.2, p. 388, The
evolutionary integration of the genetic material into genetic systems.
In Brink, R. G. (editor), *Heritage from Mendel*. Copyright 1967, University of Wisconsin Press, Madison, Wisconsin. With permission.

of making cells, Sonneborn says, "Strong direct evidence now confirms the old dictum that only a cell can make a cell" (p. 202).

We call attention to the figures of paramecium surface areas (Sonneborn's 1967 report, Fig. 19.2, p. 388, and the diagram of testate *Euglypha*, Fig. 19.3, p. 389). See our Figs. 4.2 and 4.3.

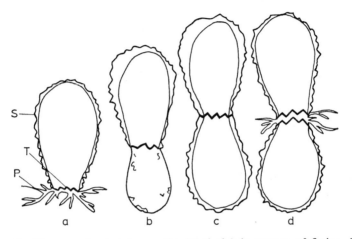

FIG. 4.3. Diagram of a testate amoeba (*Euglypha*) in process of fission. The letters indicate the following: a, b, c, d, successive stages in fission; P, pseudopodia; S, shell; T, teeth. Reproduced from Sonneborn, T. M., Fig. 19.3, p. 389, The evolutionary integration of the genetic material into genetic systems. In Brink, R. G. (editor), *Heritage from Mendel.* Copyright 1967, University of Wisconsin Press, Madison, Wisconsin. With permission.

In the Hydrozoa and Scyphozoa, however, differences do arise during asexual reproduction as we have pointed out in the previous chapter, and this occurs regularly in spite of the transmission of certain basic aspects of differentiation such as the germ layers. Here, therefore more of "development" persists, though much less than is characteristic of sexual reproduction in these forms or in other Metazoa. But the maximum amount of development is reserved for the higher organisms which reproduce only sexually. "This kind of mechanism, which is in itself most complex when one considers all the details of the nuclear and chromosomal behaviour, must have had a long, slow evolutionary history. It apparently is efficient judging from its ubiquity in most animals and plants, and furthermore its very complexity would argue against radical changes appearing, which would further account for the uniformity of the nuclear mechanisms in all organisms. But the important point for our argument here is that the processes of fertilization and meiosis occur in unitary nuclei.

"This simple fact that the mechanism which produces and transmits variation uses single cells is vital to the origin of development. Because the increase in size has, under certain conditions, adaptive value, there has been a repeated and persistent trend towards a multicellular condition. If the organism is to be multicellular and its variation system unicellular, the direct consequence will be a development. To put the matter succinctly, development is the result of sex and size" (Bonner, 1958, p. 11*).

Cytoplasmic organization in sexual reproduction

In the light of these remarks we will turn to the problems of development associated with sexual reproduction, but not expecting that only one type of development will occur in all many-celled organisms. As a matter of fact diversity of development is marked even within the phylum Cnidaria itself and has been commented upon by several (Hand, 1959; Greenberg, 1959). These facts certainly have a bearing on the selection of the criteria for the recognition of homology and the definition of the concept as we shall see subsequently.

Knowledge of the ultrastructure of the cytoplasm may be expected ultimately to show us how development is related to "sex and size" and how the structure and relative positions of homologous organs are established. We have no complete story as yet but facts are accumulating which inform us of at least some parts of the machinery involved, and these facts have led to some working hypothesis which will likely lead to the discovery of the remaining information required to define the roles of genes and cytoplasm in character determination. The available information appears to support the conclusions of Danielli (1959). He has given us the following general suggestions which are stated below in part.

* From Bonner, *op. cit.*

1. That the nucleus provides the information required for the synthesis of specific macromolecules and that it normally exercises exclusive control in this respect.

2. That the cytoplasm provides the information necessary for the assembly of macromolecules into specific structures and normally exercises exclusive control in this respect.

3. In the normal dynamic relationship between nucleus and cytoplasm the control of a character will be exerted through messenger molecules able to move between nucleus and cytoplasm and between different cytoplasmic units.

In the further discussion of these hypotheses Danielli stresses the probable importance of permeases in the movements of the messenger molecules through the cytoplasm. "Probably the greatest achievement to date of electron microscopy has been to reveal the extent to which cytoplasm is cut up into distinct regions by paucimolecular membranes." On the other hand there seem to be no such membranous hindrances to the movement of molecules within the nucleus. There are in the cyptoplasm then, many aspects of structural organization, and the "profound reorganization" following fertilization to which Danielli previously referred could do no more than reduce these to a minimum. Of organized structures there are at least the mitochondria, the Golgi region, the microsomes or ribosomes, and the endoplasmic reticulum or ergastoplasm. For the medium through which the transmission of messenger molecules can take place there is the endolymph. Evidences of persisting polarity in eggs have withstood the attacks of generations of experimental embryologists and cytologists and it seems safe to reaffirm that the egg cytoplasm normally possesses a minimum cytoplasm framework for polarity, in relation to which homologous structures ultimately develop. Full details are lacking as to the critical steps in the development of homologous structures but it is clear that development is "only conceivable in terms of a partially preformationist theory". Given this minimum organization the theory of Weiss (1962) may account for the positioning of the protein and other macromolecules just as bricks and stones and cement are added to the framework of a modern building. The marked distinction between building the framework and filling it in is one of the major results of the experimental studies of Danielli (1959).

So much for the framework; now what about the substances, the macromolecules themselves? The evidence from the study of chemical genetics indicates that these are under the control of the nucleus. The manner of their inheritance is known for a great number of such traits and the information is presented in recent texts and reports (Haldane, 1954; Wagner and Mitchell, 1955; McElroy and Glass, eds., 1957). Much study has also concerned itself with DNA and RNA and their roles in the determination of biochemical characters. A recent review (Fraser, 1959) summarizes much of the information which bears upon the possible ways in which the genes determine the nature of specific proteins during development. Here too an interaction of genes and cytoplasm is postulated even though the genes may be considered in a sense to be the primary determiners, and it is the developing *cell* that is the complete unit. DNA and RNA are the templates which provide the moulds for the construction of the macromolecules of protein. "Structural modification, the overt signal of differentiation, can only be manifestations of the handwork of proteins, because of the importance of proteins in structure at all levels."

There is also the recent account which carries the theory of character determination further, explaining and summarizing the available information. This is the report "The Living Cell" (Brachet, 1961) which covers many of the aspects of cell structure and function. Two more of the chapters in this issue of the *Scientific American** have a special bearing on problems related to the determination of homologous characters, viz., the chapter by Allfrey and Mirsky, "How Cells Make Molecules" and the chapter by Fischberg and Blackler, "How Cells Specialize".

The first of these chapters deals with the mechanism of heredity itself, that is with the replication of DNA, and with the production, under the determining influences of DNA, of RNA and in turn of proteins synthesized under the influence of messenger and transfer RNA and template RNA. A detailed account of the series of steps involved in the synthesis of a protein is tentatively outlined beginning with the synthesis of RNA by its complementary DNA, followed by the transfer of the appropriate amino acids to the template RNA in the ribosomes. The synthesis of the protein proceeds zipperwise following the selection and alignment of the appropriate amino acids.

* September 1961, Vol. 205, No. 3.

It seems clear that DNA is the hereditary code for proteins and if a cell or an organism was just a bag of proteins we could feel that the problems of hereditary determination were solved. But no cell is just a bag of proteins—much less any organism! The problems of the structural organization of cells, tissues and organs still require solution and the mechanisms of homology will not be completely understood until they are.

The chapter by Fischberg and Blackler deals with such problems and focusses attention on the organization of the cytoplasm. The early polarity of the egg and an early predetermination of the major features of structural organization are emphasized. It is suggested, however, that there is an increasing specialization of nuclei as well as cytoplasms during development, but the more recent studies of Gurdon and others make this doubtful.

It was our good fortune to be taught by M. F. Guyer at the University of Wisconsin that "the cytoplasm is not negligible in inheritance". Though at that time many decades ago the DNA story was unknown, we believe it is still a valid statement. The DNA may code only for proteins, structural and enzymic, and if so life is much ado about proteins. But it is not therefore proven that the nuclear organization is the only continuing organization from generation to generation. Heredity is the transmission of organized protoplasm from generation to generation, with its innate capacities. The cytoplasmic organization may not be as elaborate and important as Michaelis (1954) has claimed, but it would be quite incorrect to claim that "the total genetic potential of an organism is represented in DNA" (Hoyer, McCarthy and Bolton, 1964). Rather, Sonneborn (1963) is nearer the truth when he stated, "Strong direct evidence now confirms the dictum that only a cell can make a cell." And as Gurdon (1969, p. 225) has put it, "There is extensive evidence of a theoretical and experimental kind that the distribution and nature of egg components bring about the first differences between cells in early animal development."*

Though we have not yet achieved a complete and detailed description

* From Gurdon, J. B., "The importance of egg cytoplasm for the control of RNA and DNA synthesis in early amphibian development". In Hanly, E. W., editor, *Problems in Biology: RNA in Development.* Copyright, University of Utah Press, Salt Lake City. With permission.

of the mechanism of homology in any single case of sexual reproduction, we have made some progress towards this goal. Obviously
where the mechanism is simplest, as in some of the somatic characteristics transmitted in asexual reproduction, where already formed
structures are distributed among the "offspring", the mechanism is most
easily understood. Thus the cilia, the gullet, parts of the oral groove
and some of the contractile vacuoles of daughter Paramecia are
homologous (essentially similar) to those of the parent, because they
are literally the same structures! But new gullets, contractile vacuoles
and even new cilia must appear *de novo* so that ultimately each
daughter cell will possess its normal complement. And the tentacles
and hypostome of a Hydra bud must arise *de novo* even though its
germ layers and gastrovascular cavity are continuous at first with those
of its parent. Of course, our understanding is still incomplete for most
characters, though the genes, the DNA and RNA, the mitochondria,
the ribosomes and the ergastoplasm are all known to be concerned and
theories in regard to their particular roles have been developed. As
Picken (1960, pp. 503–504) says, "development is only conceivable in
terms of a partially preformationist theory. We have to project attrac-
tions and repulsions back on to the constituent molecules of the egg
in accordance with the observed regionally specific segregation of
materials. The chemical properties that Weiss invokes are familiar
ones; there is no room for a Maxwellian demon. But none the less, for
all its molecular terminology, the theory accepts as given what it is
there to explain: namely, that it is a hen's egg which develops into a
hen. It postulates a minimal, specific, spatially organized material
system which for all its minimal character is yet capable of self-ordering
itself to a fowl rather than a duck, a bird rather than a mammal, a
vertebrate rather than an invertebrate."*

   And yet our excursion into mechanisms of homology is not without
its values for the very limitations and inadequacies of our knowledge
should guide us in regard to the ways we define and discuss homology.
All the detailed knowledge of molecular organization which can be
obtained does not lessen in any way the existence of gross structural
characters or the need for their comparison. But we should avoid

   * From Picken, L., *The Organization of Cells and other Organisms*. Copyright
1960, Oxford University Press, London. With permission.

defining homology in such a way that the specifications cannot be met because of the limitations of our knowledge. Homology is indeed "morphology's central conception" and we must use the term or at least deal with the relations among structures to which it has been applied. We will consider in the next chapter what recommendations can now be made in regard to the terms homology and analogy in modern zoology.

CHAPTER 5

IMPLICATIONS OF HOMOLOGY AND RECOMMENDATIONS IN REGARD TO USES

Alternative views in regard to homology

OWEN (1866) has stated that he believed that resemblances in specially homologous series were "no doubt due to inheritance" and this obvious conclusion would lead us to expect that geneticists would have been aware of their own obligation to contribute to the elucidation of the mechanism of homology. My astonishment was very great, therefore, in finding that prior to 1932 geneticists had seldom discussed the homology of organs and parts of the bodies of organisms and when they did publish their conclusions they were completely false! Thus Duerden (1923-4) wrote:

"Following the older methods of comparative morphology, we can maintain that the relationship of scale and feather affords evidence that in the course of evolution the reptilian scale has grown upwards into a filament, and by a complicated system of incisions of the epidermis, due to ingrowths of the dermis, the filament has frayed out, and given rise to the many structural divisions of the feather—shaft, barbs and barbules. This is the view which has hitherto been largely accepted. . . ."

"But no," says the Mendelian, "a feather is a new structure; it is *sui generis*: it is an epidermal mutation, the result of a separate germinal change; its origin is quite apart from that of scales. . . ."

The delight of the morphologist in the tracing of corresponding structure through all its many transitional stages, from its one extreme to the other, is but a delusion if a genetic relationship is the underlying idea.

It is difficult to understand how a geneticist, or anyone, could be so misled in regard to the implications of genetic facts, for there were plenty of facts known at that time which would have led to a correct conclusion. The clue as to the cause of the misunderstanding can be found in the brief discussion of Crew (1925). It was his belief that a feather of a bird is not necessarily the transformed scale of a reptile but may be a distinctly different characterization based upon an entirely different genotype. A given genotype results in a certain characterization—scales; mutation—alteration—in this genotype results in a new genotype and thus expresses itself as another characterization. The old genotype is transformed into the new but the old characterization disappears and is replaced. Scales and feather, therefore, are not homologous structures—homology attempts to establish a similarity in origin and nature of structures seemingly different and is based on the assumption that during the course of evolution structures have undergone transformation yet remain fundamentally the same. In fact, this conception of homologous structures cannot be accommodated by the chromosome theory until it can be demonstrated that the genes themselves can pass through a process of gradual modification.

The chief basis for the erroneous conclusions reached by Duerden, and by Crew, are these:

1. The assumption that all mutations in genes must necessarily alter the fundamental structure of the organs conditioned by those genes.

2. The assumption that the somatic discontinuity of successive generations of organs makes the homology of organs impossible.

3. The assumption that all genes must pass through a process of gradual modification if the homology of the corresponding organs concerned is to remain.

Every one of the above assumptions is untrue, and the conclusion that genetic homology is a delusion is equally untrue. There can be no doubt in this matter; essential structural correspondences, as all characters, are determined primarily by inheritance. But, granted that this is the case, are we justified in: (1) inferring that all essentially similar characters are due to *common* heredity, i.e., *common* ancestry, and (2) defining the relationship of homology as "any structural similarity due to common ancestry". It seems to me that we cannot justify either of these practices because they cannot stand up against searching criticism. Though Simpson (1959) has presented an able defence of these usages which, it must be granted, are the prevalent ones, we challenge their validity on several grounds. In general the same criticism can be levelled at the definition of homology as "due to common ancestry" as was held against current theories of development, viz., it presumes to be true what it is attempting to prove. More particularly Simpson (1959) has attempted to make the case that Darwin implicitly provided a "theoretically objective criterion as to what is and what is not a homologue. The common ancestry that not only explains but in this view also defines homologues did objectively exist if evolution is true." Simpson goes on to say that "what constitutes 'essential structural similarity' can only be subjective or intuitive. It is not even an opinion *about* a definable reality, which judgment of common ancestry is, at worst."

We have briefly commented (Boyden, 1960) on these claims, which we consider to be without practical applications. Our rebuttal follows:

1. Inferences regarding ancestry must, in the absence of pedigree records, be drawn mainly from an appraisal of the amounts and kinds of similarity in the structures compared whether they are those of existing organisms or fossils or both. If, as Simpson says, this appraisal "can only be subjective or intuitive" it cannot then suddenly become objective merely by assuming the common ancestry and then calling the structures homologous.

2. Morphological comparisons can be made as objectively as any human descriptions can be. If it may be true in a sense that what constitutes "essential structural similarity" is subjective, it is also true that an accurate description of the structural characters themselves need not be coloured by subjective interpretations, nor is it so imbedded in assumption as judgment of common ancestry always is, even though an abundant fossil record is available.

3. The very same kinds of structures may be due to common ancestry if they are to be found in different genetically related organisms (special homologues) or not if they are to be found in different regions of the same individual organism (serial homologues).

4. Knowledge of the mechanisms of heredity is very incomplete in most organisms. In fact such knowledge is available in amounts sufficient to justify inferences regarding the hereditary basis for character determination in perhaps only a few members of four of the animal phyla, viz., Protozoa, Mollusca, Arthropoda, and Chordata. In all

other phyla, the inferences that particular characters are due to heredity have the most tentative and general basis. It seems to be thought that because they are there, these characters must be due to common heredity, which again is presuming what it is necessary to prove.

We think it is appropriate, with all due respect to the able presentation of the case for evolutionary taxonomy by Simpson (1961), to cease this beating about the evolutionary bush (or phylogenetic tree) and to attend to the description of the structures of organisms as they are. Following this description there should be comparison, and there may be genetic analysis for existing organisms. But the terms used in referring to structural agreements should be distinct from those based on unprovable inferences in regard to these agreements. A more detailed discussion in regard to these recommendations will follow.

Owen's criteria of homology

The criteria used by Owen in the recognition of homologous structures have been referred to in the previous chapter. They included:

1. Similarities in the relative position and connections of the parts.
2. Similarities in the adult structure and in the development of the parts.

The relative importance which Owen (1848) accorded these similarities can be determined from his own words.

"There exists doubtless a close general resemblance in the mode of development of homologous parts; but this is subject to modification, like the forms, proportions, functions and very substance of such parts without these essential homological relationships being thereby obliterated. These relationships are mainly, if not wholly, determined by the relative position and connection of the parts, and may exist independently of form, proportion, substance, function and similarity of development. But the connections must be sought for at every period of development, and the changes of relative position, if any, during growth, must be compared with the connections which the part presents in the classes where vegetative repetition is greatest and adaptive modification least."

In his further explanation, Owen distinguishes among several kinds of homology.

"Relations of homology are of three kinds: the first is that

above defined, viz., the correspondency of a part or organ, determined by its relative position and connections, with a part or organ
in a different animal; the determination of which homology indicates
that such animals are constructed on a common type. . . ." (Special
homology.)

"A higher relation of homology is that in which a part or series
of parts stands to the fundamental or general type, and its enunciation involves and implies a knowledge of the type on which a natural
group of animals, the vertebrate for example, is constructed."
(General homology.)

Finally, Owen refers to similar repeated parts in a single individual,
parts which he calls *homotypes* so as to be sure to distinguish them
from special homologues. In this he is much more sound that some
later authors (Hubbs, 1944; Moment, 1945) who have made the
erroneous claim that serial and special homologues have the same
meaning for the biologist! I am glad to say that Simpson (1959) rejects
these claims, which were previously criticized (Boyden, 1947).

In a footnote (1848, p. 16) Owen stated:

"It will, of course, be obvious that the humerus is not 'the same
bone' as the femur of the same individual in the same sense in which
the humerus of one individual or species is said to be 'the same bone'
as the humerus of another individual or species. In the instance of
serial homology above-cited, the femur, though repeating in its
segment the humerus in the more advanced segment, is not its
namesake or 'homologue'. I have proposed, therefore, to call the
bones so related serially in the same skeleton 'homotypes', and to
restrict the term 'homologue' to the corresponding bones in different
individuals or species, which bones bear, or ought to bear, the same
names."

In our own discussions of homology and related topics (Boyden,
1943a, 1947; Blackwelder and Boyden, 1952) we have spoken of
"essential similarities" in the structures of corresponding parts or the
"essential nature" of the organisms possessing those parts. Owen used
these terms also, but in a more abstract sense than we. Thus Owen
(1848) explained that he was using the term essential nature in the
sense of its 'Bedeutung' "as signifying that essential character of a part
which belongs to it in its relation to a predetermined pattern, answering

to the 'idea' of the Archetypal World in the Platonic cosmogony. . . . " With this philosophical abstraction we have nothing to do—when we speak of the "essential nature" of organisms we mean the characters they possess and when we speak of "essential similarities" in such characters we mean that they agree in the possession of objective criteria which can be recognized, described, weighed, measured, photographed. The objective criteria of homology are those Owen recognized as establishing the relationship of special homology (or at times of serial homology) and it is the evaluation of these criteria which Simpson says (incorrectly we believe) can only be subjective or intuitive.

Essential similarities in structural characters which would qualify them for characterization as homologues

In the first place, there are a considerable number of major characters and a host of minor characters which animals belonging to particular taxa possess. The nature of the associations among these characters establish certain *types* to which animals belong, types which have nothing to do with any abstractions of a philosophical or subjective nature. Certain of these characters are listed in the following series:

1. Organisms non-cellular or cellular.
2. If cellular, may be unicellular or multicellular.
3. Asymmetrical or symmetrical; if symmetrical, it may be spherical, radial, biradial or bilateral.
4. Number of germ layers two or three.
5. Presence or absence of tissues and/or organs.
6. Skeleton present or absent; when present, it may be internal or external or both.
7. Presence or absence of a gut.
8. Presence or absence of a coelome.
9. Body metameric or non-metameric.
10. Appendages present or absent; if present, jointed or non-jointed; unpaired or paired.
11. Nervous system present or absent; if present, may be nerve net, ladder type, or tubular. Sense organs varied.

12. Circulatory system present or absent; if present, open or closed. Hearts tubular or chambered.

13. Excretory system present or absent; if present, it may possess flame cells, or flame bulbs; nephridia or kidneys (green glands, Malpighian tubules).

15. Respiratory system present or absent; if present, may include gills, book gills or book lungs, tracheal tubes, lungs.

16. Presence or absence of distinctive types of organelles, cells, organs, e.g. cilia, flagella, cnidoblasts, simple or compound eyes, statoblasts, tube feet, Aristotle's lantern, etc.

No one can justly claim that the recognition of these characters is mainly subjective or intuitive, and, if not, their bearing upon special homology can also be recognized in most cases objectively. But, in view of the controversy regarding the cellularity of the Protozoa, it may be necessary to point out the difference between *definitions* of structures or relations and the criteria required to satisfy the definition. Thus in the controversy just referred to, it was agreed by most of those on opposing sides that the same fundamental type of protoplasmic organization was to be found in Protozoa as in the cells of Metazoa. The difficulty was that both Dobell and Hyman limited the use of the term "cell" to multicellular organisms by their arbitrary definitions. It was not at all a question in regard to the nature of the fundamental organization (except possibly in the case of the Mycetozoa as raised by Martin, 1957).

Adult versus developmental structural correspondences

There has been, and still is, a considerable difference of opinion among naturalists in regard to the relative importance of adult and developmental structural expressions as criteria indicating a relationship of homology. This is true whether the concept of homology was interpreted as indicating an agreement in structural relations merely or evidence of common ancestry. We have seen how Goethe used the skull of an immature human to solve the problem of the apparent disappearance of the intermaxillary bone in the adult. We have noted also that Owen tended to minimize the importance of the developmental stages. On the contrary, Darwin in the latest (sixth; 1872, p. 253) edition of *The Origin*, called attention to the importance of the embryonic stages in problems of the analysis of racial descent, admitting, however, that modifications of embryonic development might obscure this evidence. "Thus, community in embryonic structure reveals community of descent; but dissimilarity in embryonic development does not prove discommunity of descent. . . ."

Haeckel at first used homology with Owen's meaning but later (1866) accepted the ancestral criterion in that he stated that a true homology can be established only between parts which have sprung from the same original source and have differentiated themselves in the course of time.

Huxley (1869, p. 137) considered the embryonic criteria to be the decisive ones in the relationship of homology, for he defined it as follows: "Homology, the relation between parts which are developed out of the same embryonic structures. . . ."

Gegenbaur (1878) changed his early views in order to adopt the ancestral criterion of homology while emphasizing the importance of embryological correspondences. Thus he wrote (p. 64), special homology "is the name we give to the relations which obtain between two organs which have had a common origin, and which accordingly have also a common embryonic history. As exact proofs of genetic relations are necessary for the investigation of special homologies, this mode of comparison is generally limited in the lower divisions of the Animal Kingdom to systems of organs; it is only in the Vertebrata that it is possible to extend this method to more minute features." And in another place Gegenbaur stated (p. 63), "Blood-relationships became dubious exactly in proportion as the proof of homologies is uncertain."

The above passages are somewhat contradictory. In one place we are told that exact proofs of genetic relations are necessary for the investigation of special homologies; and in another, we are informed that blood-relationship, that is *genetic relationship*, becomes dubious exactly in proportion as the proof of homologies is uncertain. This circular reasoning remains to this day in the current definitions of special homology, a striking witness to the persistent effects of the backward look in the appraisal of morphological correspondence.

On the other hand, Wilson (1896) cites many cases where diverse developmental processes lead to essentially similar adult end results.

One interesting case is the contrast between egg and bud development in certain Tunicates as concerned in the development of the atrial chambers.

Other cases cited deal with cell-lineage studies and modes of cleavage. Sometimes these appear to be conservative, at other times not. "The puzzling facts reviewed in the foregoing brief survey leave no escape from the conclusion that embryological development does not in itself afford at present any absolute criterion whatever for the determination of homology. Homology is not established through precise equivalence of origin, nor is it excluded by total divergence. . . ." "The very statement that homologous parts differ in embryological origin itself implies some higher standard of homology that outweighs that of development." And in answer to the implied question, what is this higher standard, Wilson stated, "Obviously it is the standard of comparative anatomy."

It is true that later Wilson modified his position somewhat, expressing the hope that the study of the cleavage stages may prove as valuable a means for the investigation of homologies and of animal relationships as that of the embryonic and larval stages. At the same time, he admitted that nearly related forms, for example the gasteropods and cephalopods, may differ widely in the form of cleavage.

Spemann (1915) has contributed a scholarly review of the history and implications of the concept of homology. In this account, he states that only parts of such organisms as are built on the same fundamental body plan can be homologous, that is one seeks for homologous organs only in animals which have other obvious homologous organs. This, of course, leads directly to the idea of the type as a basic form of organization.

Spemann points out that the concept of homology has had a geometric content in that corresponding parts of *similar figures* only were considered to be homologous, and he emphasizes the fact that the geometric figure is itself unchanging whereas an organism goes through a period of development. Then what of the homologous parts—if the adult parts are homologous, are their anlage also homologous? The answer is yes, this is true in the great majority of cases. Furthermore, it is sometimes easier to trace the homologies in the early stages than in the adult ones. This certainly is true for the notochord in Chordates, also for their gill clefts. Nevertheless, Spemann comments that comparative anatomy must always remain an essentially hypothetical science to the extent that it assumes common ancestry because of common form. To Spemann it seemed more important to properly establish and assemble the criteria which indicate homology than to fill in a few gaps in homologous series or even to demonstrate new homologies. He points out that even the conservative criterion of relative position and connections is not of universal validity, for example, muscles can wander and acquire new origins and insertions. Though usually the original nerve connections remain and thus identify the muscle, even this principle is not of universal validity.

It is natural that Spemann should relate the findings of experimental embryology to the problems of determining homologies. He speaks of experiments in which the lens of the eye of a salamander may be regenerated from the edge of the iris and not from the ectoderm, its original source. However, no one would be able to distinguish the two kinds of lens if he did not know what actually happened. In his further discussion, Spemann refers to Lankester's (1870) distinction between homologous and homoplastic correspondences and points out that the distinction is very often difficult to make.

Finally, Spemann states that in spite of these difficulties with the ancestral criterion of homology, he will not completely throw away the concept of homology in the historical sense. On the contrary he looks upon organisms as historic beings and not merely as so many variations on an ideal type. All that can be learned of their development, he says, is useful knowledge—only we will not believe any more that we first establish the phylogenetic tree and then determine the laws of development. It is the other way around and thus the goals of morphology have changed from those of the early post-Darwinian period.

Naef (1926 and earlier) has critically discussed many aspects of the concept of homology, emphasizing its importance in relation to systematics rather than as a basis for the study of phylogeny or analytic embryology. He states that homology may only be spoken of in relation to type similarity, but that evolution has provided the explanation for such similarities. However, since the developmental stages of any organism are concerned with the evidence of the type, they become important evidences of homology and may be at times more conservative than the adult expressions. A rule regarding the designation of developmental correspondences is stated, "Parts which are known to be homologous in one stage of ontogeny remain homologous in the course of all individual changes, so long as they remain separate from others before and behind." This rule is a logical addition to our purely formal expression of the homology concept, but must not be interpreted to mean that generally the embryonic stages are more important than the adult stages in the recognition of homologies. It could, for example, not be applied to cells in an indifferent or undetermined stage, even though later they are determined and become particular parts of the adult expression. Naef does not view favourably the attempts to further subdivide the concepts of special and general homology, for the more precise expressions may transcend the limits of the available knowledge. Parts are either homologous or non-homologous, even though their expressions may run through many gradations. With this approach we avoid the unnecessary confusion in regard to the determination of the special homologies of repeated structures, such as a particular vertebra or a particular tooth, determined by number from some point of reference, e.g. the atlas in the case of a vertebra, or the canine in the case of a tooth. It is well known that the sacral vertebra may have a most varied numerical position in vertebrates, but in spite of this, it always has the same essential relation to the pelvic girdle and all such sacral vertebrae should therefore be considered homologous. Trunk vertebrae of one vertebrate are homologous with trunk vertebrae of another, even though there may be several times as many in one organism as in another. To attempt to homologize them by number is to give them an historic and ontogenetic individuality which they do not really possess!

In concluding his report, Naef has said that the divergent ideas and perplexities reported over the last century and a quarter have confirmed his statement of 1913: "I see therein an expression of the slight heeds of logical clarity which characterize most biologists; we operate continually with entirely hazy concepts and base upon them the longest debates. Only with logic and more critical concepts and laws may we make progress." These words do not mean that Naef does not recognize the phylogenetic approach to the study of life, but that he considers it to be based upon the facts of "idealistic" morphology. He finds only a bare tautology in the definition of homology as resemblance due to common ancestry, and not a new criterion for the recognition of the relationship of homology.

It would be expected that the idea of recapitulation should naturally be discussed here, but we have already dealt with this topic in Chapter 3. Hence we will conclude our discussion of adult versus developmental characters with a brief mention of what has been called heteromorphosis in which an organ or appendage of a different type is regenerated at the site of its predecessor. Does this imply that the heteromorphic organ is homologous with the normal organ which preceded it? According to a strict application of the principle of connections we should conclude that an antenna, regenerated at the site of a stalked eye in Decapod Crustacea is homologous with the eye. It is doubtful, however, that this would be a useful conclusion to draw on either the morphologic or ancestral criteria of homology. The regenerated antenna is no more similar to the stalked eye than the normal antenna, the two kinds of appendage being structurally very different. We believe it would serve the study of biology best to realize that the cells which normally make an eye in these organisms had also originally competence to make antennae and that this competence is retained or regained by the cells at the base of the amputated eye stalk which regenerate the antenna under the experimental conditions. There must be essential structural similarity as well as agreement in relative position and connections in order to satisfy the specifications for the relationship of homology.

In view of the divergent views and varying developmental phenomena briefly reviewed the only conclusion that seems warranted is that no general rule can be stated regarding the relative importance of developmental and adult structural characters in the determination of special homology. Whether this homology is a morphological concept only, or is defined in terms of common ancestry, each group must be studied in its own right, and the relative consistency and weight to be assigned to the developmental and adult characters determined by the results of such comparative study. Not enough is as yet known about the molecular detail of the mechanism of homology except for proteins to provide certain criteria for the relationship implied in "morphology's central conception". It would be the better part of wisdom, therefore, to bear this in mind when definitions of homology are considered, and not to define the concept in terms of specifications which cannot honestly be met.

Morphology and typology

In these times it seems that to be a "typologist" is to be a living anachronism and such a person is likely to be ridiculed and scorned. One must not speak of animal types, nor even of kinds of organisms; all biological understanding is now said to relate to "populations" and common ancestry. But we ask, populations of what? And what *is* a population? Those who use these terms seldom bother to define them and we are left with the same objective realities as before, viz., individuals of many kinds, with varying amounts and kinds of similarity. To bring order to the study of these individuals, requires first that they must be described and compared; then named and grouped in accordance with the available knowledge of their characters. In the process of doing this it is discovered that there are characters and associations of characters by means of which the individual organisms can be grouped into "natural systems". It is our contention that the most natural of all taxonomic systems would be those in which organisms are grouped in accordance with their natures; that is in accordance with their characters and associations of characters.

It was perfectly natural then that early in the study of animals, types of organization became apparent. These structural plans were, in the pre-evolutionary days, interpreted differently from now. Sometimes they were considered as "plans of creation", which were assembled into various fanciful "natural systems", e.g. the quinary system of Vigors (1825), MacLeay (1825) and Swainson (1835). Strickland (1844, 1846) criticized these arbitrary arrangements and attempted to group animals in accordance with their degrees of "affinity", that is, in agreement with the amounts of similarity in their more conservative characters. At this time "affinity" had no implications in regard to genetic relationships, but serious efforts were made to distinguish relationships of "affinity" from those of "analogy", the latter having to do with the more superficial characteristics of the organisms. In a sense Strickland was more scientific than many present-day zoologists for he was more interested in the natures of the organisms studied than in speculations in regard to arbitrary systems or remote and unprovable ancestries.

The coming of Darwinism brought mixed blessings. First there was the result that natural history could become more natural and less supernatural, since a natural basis for the origins of species was provided. But, this benefit was unfortunately to be partly counteracted by too much emphasis on ancestry as subject for study; on the backward look, leading to the most unscientific attitudes in regard to taxonomic work. A host of phylogenies grew up like weeds in the garden of systematics, and even organisms with no fossils available for study were claimed to be classified in accordance with their phylogenies! Another unfortunate result was that many able biologists turned away from such an unscientific approach to systematics and comparative study suffered in consequence.

We need to attend to the natures of organisms as revealed in their structures and compositions and functions. In fact, the most important thing to know about organisms is what kinds they are—this is the first and the greatest knowledge without which no systematic analysis is possible. And in assembling this knowledge information in regard to the types of structure will become available. This is the new objective typology, a branch of the study of the organization of matter, now being carried on by the newer methods for the study of fine structure, as well as by the study of gross anatomical structure. But the latter remains an essential part of morphology and it might even be true some day that to be a comparative morphologist, or even a typologist, will be more commendable than to be a guesswork phylogenist. It is sometimes difficult in these times to understand the extreme criticisms of "typological thinking" made by Kiriakoff (1959), Simpson (1961), Mayr (1965) and others.

According to Mayr (1965), "The real trouble with typology in the philosophical, conceptual sense, is not that it is merely 'contaminated with idealistic, metaphysical concepts', but that it is strictly and exclusively based on such concepts, that it is an incarnation of such concepts and simply cannot be separated from them" (p. 91).

E

The trouble with Mayr is that he is wrong when he says that a consideration of animal types cannot be separated from the mysticism of the old "plans of creation". Charles Darwin did it neatly and completely in his *Monographs on the Cirripedia* (see our Chapter 7). Another complaint about Mayr is that he seems unaware that the meaning of such terms as typology can change and have done so in modern times. That meanings of even very common biological terms can change Mayr should be aware of, for he, following Bock (1963), is attempting to completely eradicate the original meaning of homology as essential structural correspondence and replace it with the meaning of common ancestry without any requirement of correspondence, structural or otherwise.

How many animal types?

Opinions in regard to the number of animal types have varied and may continue to do so. It is the same with regard to the number of phyla, or any other systematic category. But the structures and structural organizations themselves are relatively objective; it is only the limits of inclusiveness concerning which there is difference of opinion. It is the same with the related concept of homology; where is its endpoint? Are the alimentary tracts of all animals homologous in Owen's sense? Their characteristics are objective but the degree of similarity is relative. Sometimes, in fact, the relative amounts of similarity can be measured but even in such cases an arbitrary endpoint may have to be fixed where the similarities are continuously graded. On the contrary, a type may be fixed by the presence or absence of some particular structure such as a vertebra, in which case that particular type has been given the taxonomic rank of subphylum in most current classifications. Of course, there are many associated structures in the organization of every type and in the case of the Vertebrata, these associated structures extend to and delimit the phylum Chordata. Types may then at times agree with phyla in their inclusiveness, or they may extend beyond a single phylum. Varied as they are in limits, the structural characters upon which they are based can and should be objectively determined. And likewise the determination of homologies becomes the more objective, the more it is based on the structures themselves and the less on phylogenetic inferences.

Different views in regard to animal types have been held by our greatest zoologists, and a review of the changes in animal classification would reveal some of these views. We do not intend to include in this account the "types" recognized by the German naturphilosophers, such as by Oken, or those based on any other criteria except their structural characters and associations of characters.

Cuvier (1817) recognized only four plans of structural organization; rejecting any "scales of nature" such as described by Bonnet or Lamarck (1809). These four were: I, Vertebrata; II, Mollusca; III, Articulata; IV, Radiata.

The first three of these are well recognized as plans of structure today, but the last is most heterogeneous, containing a great variety of associations of structural parts and including many invertebrate phyla.

Von Baer (1828) also accepted these major types and gives some explanation as to what constitutes a type. "By *type* I mean the relative position of the organic elements and of the organs."

"The type is totally different from the grade of development, so that the same type may exist in many grades of development, and conversely, the same grade of development may be attained in many types. The product of the grade of development with the type yields those separate larger groups of animals which have been called classes."

In addition to these archetypes, Von Baer admitted that intermediate forms did occur, and that the various grades of development within the major archetypes characterized subordinate types which are called the classes of animals. And within these are to be found the sub-groups, viz., the families, genera and species. But he strongly emphasized that each of the major archetypes has its own kind of embryonic development during the course of which the embryos diverge more and more from each other. There was for Von Baer no such thing as the idea that the embryos of the higher animals pass through the permanent forms of the lower animals *en route* to adulthood, nor any uniserial arrangement of the permanent forms of animals.*

* Von Baer's views are also presented in Henfrey, A. and Huxley, T. H., editors (1853), *Scientific Memoirs, Natural History.* Taylor and Francis, London.

It was inevitable that with increasing knowledge of animal structure, the heterogeneous Radiata should be broken down into many subordinate types or groups. There is no need to detail these additions to knowledge since there is no new principle involved; it is a matter of increasing knowledge of the diversity of animal structure, obtained for the most part by objective anatomical study. We have referred to Owen as an exemplar of such methods of study but his concept of general homology was tinged with philosophy, and it was only in his dealing with special and serial homologies that Owen was relatively objective.

A better exemplar of morphological and systematic objectivity is Thomas Henry Huxley, whose *Introduction to the Classification of Animals* (1869) is most matter-of-fact. Huxley (p. 6) proceeds directly from the structures of animals to their taxanomic grouping, even though the discussion is post-Darwinian.

"Morphological classification, then, acquires its highest importance as a statement of the empirical laws of the correlation of structures; and its value is in proportion to the precision and the comprehensiveness with which those laws, the definitions of the groups adopted in the classification, are stated." He then proceeds to begin with a description of 28 classes distributed among the four Cuvieran archetypes as follows.

Radiata:

Gregarinida	Hydrozoa
Rhizopoda	Actinozoa
Radiolaria	Polyzoa
Spongida	Scolecida (?)
Infusoria	Echinodermata

Articulata:

Chaetognatha	Arachnida
Annelida	Myriapoda
Crustacea	Insecta

Mollusca:

Brachiapoda	Pulmogasteropoda
Ascidioida	Pteropoda
Lamellibranchiata	Cephalopoda
Branchiogasteropoda	

Vertebrata:

Pisces	Aves
Amphibia	Mammalia
Reptilia	

In his anatomical studies, Huxley used Owen's meanings for homology and analogy except that he considered similarity in embryonic development to be the more decisive criterion for special homology.

We conclude our discussion of the objectivity of morphological study with references to D'Arcy W. Thompson's *On Growth and Form* (1917, 1942, reprinted 1951). In his final chapter, "On the theory of transformations or the comparison of related forms", there is to be found an objective treatment of the subject. We quote several passages.

"In a very large part of morphology, our essential task lies in the comparison of related forms rather than in the precise definition of each; and the *deformation* of a complicated figure may be a phenomenon easy of comprehension, though the figure itself has to be left unanalyzed and undefined. This process of comparison, of recognizing in one form a definite permutation or *deformation* of another, apart altogether from a precise and adequate understanding of the original 'type' or standard of comparison, lies within the immediate province of mathematics, and finds its solution in the elementary use of a certain method of the mathematician. This method is the Method of Coordinates, on which is based the Theory of Transformations . . ." (p. 1032).

"We shall strictly limit ourselves to cases where the transformation necessary to effect a comparison shall be of a simple kind, and where the transformed, as well as the original, coordinates shall constitute an harmonious and more or less symmetrical system. We should fall into deserved and inevitable confusion if, whether by the mathematical or any other method, we attempted to compare organisms separated far apart in Nature and in zoological classification. We are limited, both by our method and by the whole nature of the case, to the comparison of organisms such as are manifestly related to one another and belong to the same zoological class. For it is a grave sophism, in natural history as in logic, to make a transition into another kind" (p. 1034).

Following a series of illustrations of the application of the method to the comparison of the forms of certain plants and of certain animals and parts of animals, D'Arcy Thompson concludes with the following statements:

"We may fail to find the actual lines between the vertebrate groups, but yet their resemblance and their relationship, real though indefinable, are plain to see; there are gaps between the groups, but we can see, so to speak, across the gap. On the other hand, the breach between vertebrate and invertebrate, worm and coelenterate, coelenterate and protozoan, is in each case of another order, and is so wide that we cannot see across the intervening gap at all . . ." (p. 1093).

"A 'principle of discontinuity', then, is inherent in all our classifications, whether mathematical, physical or biological. . . . In short, nature proceeds *from one type to another* among organic as well as inorganic forms; and these types vary according to their own parameters, and are defined by physico-mathematical conditions of possibility. In natural history Cuvier's 'types' may not be perfectly chosen nor numerous enough, but *types* they are; and to seek for stepping-stones across the gaps between is to seek in vain, forever.

"This is no argument against the theory of evolutionary descent. It merely states that formal resemblance, which we depend on as our trusty guide to the affinities of animals within certain bounds or grades of kinship and propinquity, ceases in certain other cases to serve us, because under certain circumstances it ceases to exist. Our geometrical analogies weigh heavily against Darwin's conception of endless small continuous variations; they help to show that discontinuous variations are a natural thing, that 'mutations'—or sudden changes, greater or less—are bound to have taken place, and new 'types' to have arisen, now and then. Our argument indicates, if it does not prove, that such mutations, occurring on a comparatively few definite lines, or plain alternatives, of physico-mathematical possibility, are likely to repeat themselves; that the 'higher' protozoa, for instance, may have sprung not from or through one another, but severally from the simpler forms; or that the worm-type, to take another example, may have come into being again and again" (pp. 1094, 1095).*

* From *On Growth and Form* by D'Arcy Wentworth Thompson, Cambridge University Press. With permission for all the quotes.

We feel it to be most refreshing to find these illustrations of greater objectivity in the study of comparative morphology. It is a greater degree of objectivity than is implied in Simpson's (1959) remarks about Darwin having provided a "theoretically objective criterion of what is and what is not a homologue. The common ancestry that not only explains but in this view also defines homologues did objectively exist if evolution is true." Let us remember that inferences regarding ancestry, in the absence of pedigree records, are based primarily on the amounts and kinds of structural correspondence; and that to be reasonably justified there must be information that the particular kinds of characters involved in the comparisons are due to the operation of hereditary mechanisms. It is acceptable to us to infer that the members of each of the types of structural organization, whether these be classes, or sometimes phyla, or even related phyla such as Annelida and Arthropoda, are genetically related, but this is an inference from the existence of obvious correspondences in structural organization—from homologies in the sense of Owen. And to be sure that we distinguish fact from theory we would continue to use "homology" in Owen's sense and to state our inferences in other and distinctive terms.

But now to return to that embarrassing question, "How many animal types?" We cannot give any definite answer to this question since the limits to a given "type" as well as to the concept of homology itself may be somewhat arbitrary. Plans of organization may be revealed in our systematic groupings, but the number of types would still be a matter of judgment. It is relatively easy to distinguish certain phyla from each other, e.g. the Ctenophora, Annelida, Arthropoda, etc., and the subphylum Vertebrata is clearly limited. But many zoologists would consider that the Annelida and Arthropoda belong to one major type, hence the limits of phyla may or may not coincide with those of the types concerned. And even the number of phyla varies in modern texts and references. Thus Pearse's *Zoological Names* (second edition, 1947) lists 28 phyla. The fourth edition (1949) lists 33 phyla. Hyman's (1940) classification lists 22 phyla. Her arrangement is based mainly on morphological structure and does succeed in expressing as well as any listing of phyla can, an interpretation of the relationships of the organisms within each phylum. By including as classes in the phylum Aschelminthes, the Rotifera, Gastrotricha, Kinorhyncha (Echinodera),

Nematoda, Nematomorpha (Gordiacea), and Acanthocephala, she has indicated that these organisms may be (with the possible exception of the Acanthocephala) more closely interrelated than is characteristic for the members of different phyla. But whether these classes thus grouped together would be said to conform to a single type is again a matter of opinion.

In sum, we suggest that a cautious morphologist may wish to limit his term homology to those animals which he considers to be of the same type. However, he may not find other morphologists in complete agreement with him, even on the same principle, for they may have a different evaluation for their types. More important than that, this procedure would, in effect, misrepresent the actual amount of structural similarity existing among organisms of different types with regard to particular structures, e.g. the cilia and flagella previously mentioned. And what harm would there be in admitting that a great many cilia and flagella are essentially similar whenever they are found, i.e. homologous in Owen's sense? Surely this is a more objective approach to the study of structure than to minimize or ignore obvious correspondences by arbitrary limitation of the concept of homology.

The concept of homology as limited to organisms of the same type

There seems to be two main tendencies in regard to the limitation of the homology concept to organisms of the same type. The pre-Darwinian morphologists and some of the post-Darwinian morphologists have generally placed definite limitations on the concept, restricting it to intra-type comparison. Thus Spemann (1915), Naef (1927), Jacobshagen (1925) and others define homology in terms of type similarities which actually brings about a closer agreement in the diagnosis of homology than would otherwise obtain. The fact is that organisms belonging to the same type of organization are more likely to be genetically related than those belonging to diverse types and this is admitted to be the case. But what this means is simply that you cannot recognize a relationship of homology between a set of the corresponding parts of two organisms unless there are other sets of corresponding parts also, among which agreement in the general plan of structure is also evident. Phylogenetically speaking, this limitation would be a valid one only if evolution has always been an intratype divergence and never has resulted in the development of one type from another. Idealistically speaking, it is a purely arbitrary limitation, which assumes that there can be no essential similarities among the parts of organisms unless there are a considerable number of such similarities.

Without in any way denying the reality of types of animal organization, e.g. the Vertebrate, the Echinoderm, the Annelid–Arthropod, etc., or claiming that all animals must be considered as genetically related by a strictly monophyletic origin, I do not believe that these arbitrary limitations are justified. The guts of animals of many types may be considered to be homologous in Owen's meaning at least from Nemathelminthes on. And surely the cilia and flagella which are constructed on the "9 + 2" basis are essentially similar in structure though they are to be found in plant and animal phyla.

Whether these appendages have such a structure "because of common ancestry" is, and probably will long remain, doubtful. It may be that broader laws of the organization of matter similar to those bearing on crystal structure are chiefly responsible. Pantin (1951) has spoken of resemblances in certain chemical substances, e.g. acetyl choline, haemoglobin, etc., the distribution of which has little apparent relationship to the genetic relationships of the organisms possessing them. Above the molecular level the materials of the nature of keratin and chitin are widely formed and utilized. The structure of striated muscle and, on a larger scale, the organization of sense organs and even of central nervous systems show similarities which seem quite unrelated to phylogenetic origins. "The organism is thus built up of standard parts with unique properties." In summarizing his discussion on organic design Pantin says, "The principles which emerge are not those which were stressed by the phylogenetic morphologists of the end of the last century. They do not contradict evolutionary principles: they are additional to them. They apply to all grades of structure, from the molecule up to the complex computing network of the central nervous system. We recognize that, as in the case of bridge building, each functional problem before the organism admits of certain possible forms of solution. To meet one or other of these solutions, structural systems can be built by utilizing the unique emergent properties of the molecules and higher orders of structural unit which are available. Natural selection will determine which of the various possibilities actually survives. It is therefore not moulding an infinitely plastic organism. It is rather directing it from one possible state to another, rather after the fashion of the moves in a game of chess."

Though Pantin does not explicitly say so, it would be natural and logical to apply Owen's definition of homology to the resemblances referred to as apparently unrelated to phylogeny which occur in diverse types of organisms. It is doubtful if anyone could deny that all cilia and flagella built on the 9 + 2 pattern are essentially similar. Grimstone (1959) explicitly recommends that the corresponding organelles of cells be considered homologous in the pre-Darwinian sense as the implications in regard to ancestry are most uncertain.

Recommendations in regard to the uses of the principal terms

Homology

Our first goal is to find that usage of these terms which is most likely to lead to mutual understanding, with due regard to the history of their development. My own views (Boyden, 1935, 1943, 1947) have led in the direction of Owen's usage for both, and I have been cited as the author of the slogan "back to Owen". No one questions the "backward look" in zoology more than I, as will be clear in what has been, and will further be, said. But in a sense it is the "confusion of tongues" that has resulted in our difficulties, for since Darwin there has developed a strange cult of ancestor worshippers, who, generally lacking understanding in the operation of genetic mechanisms and certainly incapable of ever determining this with fossil material, yet insist that we must define homology as "similarity due to ancestry" where the only available knowledge which could support this inference is the similarity itself, together with the location in time and space of the fossil structures if such are available.

The term homology has in fact been used for two main kinds of relationship! (1) Originally it was the relationship of essential similarity in structure and (2) since Darwin it has become the relationship of common ancestry. The original and still necessary meaning for homology, as it became established in the mid-nineteenth century, was essential structural similarity, and the criteria for the recognition of such a relationship were described and illustrated by Owen in 1843, 1847 and 1866 and elaborated by Remane (1956, 1963). It was a part of Owen's genius to have clearly distinguished similarity in the functions, that is the use of parts (analogies) from similarity in the structural organization of those parts, and in their relative positions in the bodies of the organisms compared. And Owen made it quite clear that homologous organs generally had the same uses but that this was not at all necessary—they might have different uses or no apparent uses at all. On this basis it was, and still is, possible to compare the structures of organisms without confusion as to the kinds of similarities concerned, though there was then, and no doubt always will be, uncertainty in regard to the final limits of the relationship defined as homology.

It is implied in defining and using homology as Owen did that (1) this is an essential concept in comparative morphology and (2) that zoologists can recognize these fundamental similarities when they see them. It seems hardly necessary to defend further the first of these two implications for the concept of homology is indeed "morphology's central conception" (Huxley in De Beer, 1928). Darwin spoke of comparative morphology as being the very soul of zoology and many others have stood in agreement in principle. The advance that Owen made was in a sense to sharpen the dissecting tools to be used by limiting the comparison to corresponding parts of the bodies of organisms, so defined chiefly on the basis of their relative positions and connections in the bodies of the organisms compared and secondarily in terms of their correspondence in adult structure and development.

But now to further evaluate Owen's direct criteria for homology we should, as Owen did, clearly distinguish between serial and special homologies, a matter of some importance because recent authors have been very confused about this. We find for example that Moment (1945) has claimed that "Bateson and Hubbs have both demonstrated that the distinction between special, serial and general homology is not a valid one". To make a long story short, I would say that Bateson and Hubbs neither have nor can demonstrate any such thing, for these different kinds of homology do have a very distinct biological significance.

It is therefore first necessary to distinguish between the two kinds of structural correspondence referred to as serial and special homologies. By serial homology we mean the essential similarity among corresponding parts distributed along a major axis of the body of *one* individual. By special homology we mean the correspondence between parts of the bodies of two or more individuals, established by the kinds of agreement in adult structure and development and in relative position and connections emphasized by Owen, and elaborated by Remane (1955, 1956) and others.

The relations between these two kinds of structural correspondence could be nicely illustrated in any metameric organism, annelid, arthropod or vertebrate. In the case of the vertebrate, taking one individual, a tetrapod, we would find serial homologies in (a) fore and hind limbs, (b) vertebrae, (c) ribs, (d) spinal nerves, etc. These same organs would also be specially homologous with their counterparts—"namesakes" in the bodies of different individuals. These structural relations can perhaps best be illustrated by Arthropoda and particularly some common Crustacea. The figures will make the comparisons graphic (Boyden, 1943).

In Fig. 5.1 five developmental stages are shown, characteristic of the individuals representing the primitive type X, and two different derived types independently descended from X, viz. Y and Z. The letters A B C D and their primes represent groups of genes which remain constant in each ontogeny, but have come to differ, in part, in the new types of descendants. The letters underlined represent groups of genes which come into action in the particular developmental stages in which they are found.

The diagram shows in a simple way the operation of the genetic mechanisms concerned in ontogeny and phylogeny. Note particularly that gene constitution stays the same during any ontogeny, but comes to differ during the succession of divergent ontogenies called phylogeny. Thus, in the case of the serial homologues within any single individual, both their resemblances and differences are a result of the interaction of the *same genes* in all cells with the various kinds of cytoplasm characteristic of that organism. But in the case of the special homo-

ONTOGENY

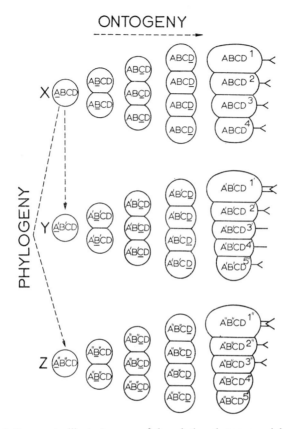

Fig. 5.1. A diagram to illustrate some of the relations between serial and special homologues and, in the larger view, relations between ontogeny and phylogeny of homologous parts. The mechanisms illustrated are those explained in part in the previous chapter. This figure is modified from Wright (1934) with permission of author and editor, and it is reproduced from Boyden, Alan (1947) Homology and analogy. A critical review of the meanings and implications of these concepts in biology. *Amer. Mid. Natur.* **37**, 648. With permission.

logues illustrated by the corresponding appendages of the adult types X, Y and Z, the resemblances are due to the same or similar genes interacting with the same or similar cytoplasms of the diverse types compared, whereas the differences are due to different genes interacting with the same or different cytoplasms characteristic of these types.

Serial homologues, being intraindividual corresponding parts, are recognized as such on the basis of their essential similarities only. Special homologues, even though their differences are products of evolution and their resemblances are due to inheritance, are also recognized as such because of their essential structural characteristics. Lacking an actual pedigree, or an almost perfect fossil record, no one could tell whether X, Y or Z was the more primitive type and thus be able to provide the correct interpretation of the phylogenetic events illustrated in the figure. Homology cannot by itself provide us with pedigree records. Assuming that X is the more primitive type, and that Y and Z are two independently evolved types derived from X, then the following processes are illustrated in Fig. 5.1:

1. Addition
   (a) of a metamere (5'; 5").
   (b) of an appendage (5').
2. Loss
   (a) of an original appendage (4").
3. Specialization of appendages
   (a) Increase in size and complexity (1—1'; 1—1").
   (b) Decrease in complexity (3—3'; 4—4').
4. Parallel evolution
   (a) In addition to metameres (5' and 5").
   (b) In specialization of parts (1—1'; 1—1").
5. Divergence
   (a) In number of appendages (5 pairs in Y; 3 pairs in Z).
   (b) In the nature of appendages.
      (i) change from biramous to uniramous (3—3'; 4—4').
      (ii) loss (4—4").

The sketch of these mechanisms shows some interesting and significant contrasts in their mode of operation, even though both deal with the interaction of nuclei and cytoplasms. First, the differences in serial homologues are not normally due to differences in genes which make up the genotype of their cells whereas differences in special homologues usually are due to just such differences. Second, the mechanism of differentiation in serial homologues is a part of the larger and more inclusive mechanism of embryologic differentiation which operates so as to produce rapidly, often within hours or days, not only differences in serial homologues but all the diverse types of structures which may arise out of a single egg, including many non-homologous structures. On the other hand, the mechanism of special homology is of such a nature as to conserve with remarkable fidelity over long periods of time involving many successive generations, those essential similarities in structure and development as well as the even more conservative agreements in relative position and connections which are so characteristic of special homologues.

Thus an analysis of the mechanisms of serial and special homology reveals their true natures and fully justifies the attempt to distinguish between them in our thinking. Serial homology arises in and is limited to ontogeny; special homology is directly concerned with taxonomy and phylogeny, and clarity and effectiveness of thought demand that we distinguish between these phenomena in our thinking and writing.

Now it is true that degrees of similarity and difference among serial homologues may appear to be comparable to the structural gradations among special homologues. Thus the third pereiopod of the lobster *Homarus americanus* is chelate and the fourth is not, while the third pereiopod of the blue crab, *Callinectes sapidus*, is also non-chelate. The extreme limits of the relationship of both kinds of homology are also the same, viz., either one can extend to the complete absence of a part. Thus in crabs the uropods and some of the other pleopods may be lacking, whereas in crayfish and lobsters they are present. Viewed from a within-individual or from a between-individual standpoint, the divergence has gone to the same total extent in either direction. But the mechanisms responsible for these similarities and divergencies are still significantly different as we have already explained and their meaning to the biologist is vastly different! No problem of systematic, or genetic, or phylogenetic relationship arises between the parts of the body of a single individual! All such parts arise from a single egg or asexual product, and their resemblances and differences have no direct bearing on the post-Darwinian concern over evidences of community of descent.

In marked contrast the comparisons among special homologues do

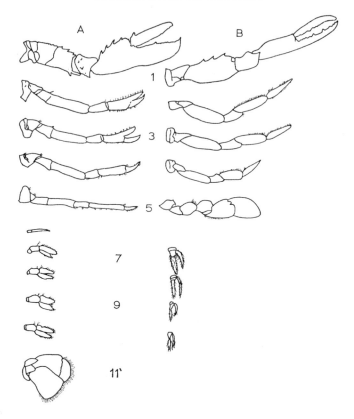

Fɪɢ. 5.2. Homotypic appendages of A, the American lobster (*Homarus americanus*) and B, the Blue crab (*Callinectes sapidus*). Special homologues are arranged at corresponding levels and numbered similarly. The pereiopods are numbered from 1 to 5 inclusive, the chelipeds being considered as first pereiopods. The pleopods are numbered from 6 to 10 inclusive, the first pair being absent from the female crab. The uropods, number 11, are also absent from both sexes of the crab. Reproduced from Boyden, Alan (1943) Homology and analogy. A century after the definitions of "Homologue" and "Analogue" of Richard Owen. *Quart. Rev. Biol.*, **18**, 238. With permission.

definitely and directly bear upon problems of systematics, evolution and phylogeny, whether the term homology is defined as resemblance due to common ancestry, or in terms of structural similarity. Which of these bases for the decision in regard to the existence of a relation of homology is the more objective? Simpson (1959) has recently reaffirmed his own view that the ancestral criterion of homology is "theoretically objective", whereas Owen's criteria of structural correspondence were stated to be subjective. Simpson goes on to say that present usage is preponderantly that of post-Darwinian naturalists wherein the relationship of homology is *defined* by common ancestry, and that there can be no sensible argument with this usage. I certainly object to this conclusion for there is a sensible argument against it. And that argument is based simply on the nature and limitations of our knowledge about ancestry, which in the absence of pedigree records, is indirect. If it is true that judgment of essential similarity in structure can only be "subjective or intuitive", then judgment of common ancestry is more so, even where a good fossil record is available because *more*, not less, intuitive judgments are required to justify: (1) a conclusion that corresponding structures (i.e. essentially similar ones) are being compared, (2) that available knowledge regarding such characters indicates that they are determined mainly by inheritance, (3) that the presumed ancestral organisms did live when and where such ancestors were assumed to be. There is *no* way to reach a scientific conclusion that organs or parts of the bodies of different organisms are homologous because they have common ancestry, except by the pathway above, and it would, as Woodger, in Le Gros Clark and Medawar (1945), has stated, be a step in advance if we would be clear in our own understanding of this basic fact and use a terminology which was in accordance with the necessary limitations of this knowledge.

Efforts to clarify the nature and uses of homology have been very numerous. One of the earliest of these, that of Lankester (1870), has a direct bearing on our present discussion. Lankester prepared to substitute for special homology as due to common ancestry, a more sharpened and more restricted term homogeny, viz., that organs or parts are homogenetic only if they can be traced to a single representative in a common ancestor. So now really, instead of common ancestry, we are to talk about a most *un*-common ancestry, that is we must go back to the *single* representative in the line of evolutionary descent, from which all later expressions have been derived! This definition was accepted by De Beer (1928) who stated that "the sole criterion necessary to establish the relationship of homology is descent from one and the same representative in a common ancestor". And Haas and Simpson (1946) who seem to regret that Lankester's term homogeny did not come into general usage, intend that *homology* should actually carry Lankester's meaning. Hubbs (1944) has accepted this but explains the great difficulties which are inherent in trying to implement it among closely related organisms, where independent modifications produce the same results in character expression as implied in the concept of homology but the time relations are different. Haas and Simpson recommend that wherever the available knowledge makes it apparently possible, that strict homogeny is the proper relationship to single out from all other similar evolutionary sequences, however close the apparent parallelism or end products obtained. Julian Huxley (in De Beer, 1928) called attention to the kind of information from modern genetics which shows how difficult a strict determination of homogeny may be. Huxley pointed out that morphology's central conception was being modified by genetics. The occurrence of independent but identical mutations in genes show that the concept of common ancestry is no longer essential to the idea of homology. Strangely, Huxley does not mention that it was not so for Owen either.

The fact is that the requirement for establishing homology in Lankester's sense of homogeny is too fine and exacting for the kinds of knowledge available *even where a good fossil record exists*! How can one ever be certain that he is dealing with a single representative of a character in the common ancestor? The establishing of the precise time relations among the generations of individuals possessing a new character would be impossible where large numbers of individuals are available as fossils, and difficult even where there are relatively few individuals, separated by appreciable and measurable thicknesses of strata. And if the fossils are not in adjacent layers of about the same age, the determination of the sequence of generations would be still more difficult. There are many ways in which known genetic processes could interfere with the correct determination of a relationship of homogeny. Among these are:

1. The separate phenotype expressions of particular genetically identical and recessive characters in different individuals. Here the particular gene mutation (or mutations) would not be expressed in the

diploid organism in which the mutation actually occurred, but only in homozygous descendants. When these homozygous individuals first appeared would depend on the system of breeding. With cross-fertilization and outbreeding the appearance of such mutant expressions could be delayed for many generations. When the system of breeding became narrower, as for example if the population were broken up into smaller inter-breeding units by isolation or extreme reduction in numbers, the mutant character could then appear and it could appear in several such populations at the same or different times! The chances of fossilization of that character in *the* common (rather the most uncommon) ancestor are infinitely small and the probabilities would rather be that "the common ancestor" would never be identifiable even if, on a remote possibility, it was actually recovered among others.

2. Recurrent mutations, i.e. the same mutations to the same gene do take place. Paleontological methods would be incapable of distinguishing among these as to which was the closer to the "single representative in a common ancestor" unless the fossil record were so poor as to space the fossils at recognizable time gaps so that a sequence could be established, but yet good enough to have a correct representation of this time sequence.

It seems to me we have enough information at hand to justify the conclusion, that the specifications for establishing the relationship of homology in the sense of Lankester's homogeny are much too exacting to be used in a precise way, even where the fossil record is good. Where there are no fossils obviously it cannot be used at all! For we cannot trace any corresponding characters to a single representative in a common ancestor if there are no known ancestors! We may have only modern descendants to deal with, none of whom is ancestral to anything but its own immediate offspring! Though a paleontologist may have overlooked this fact, it is more difficult to understand how neontologists could do so if they stopped for a moment to realize the implications of homogeny. De Beer (1958) has now achieved a more correct understanding of the limitations of our understanding of the genetic mechanism of homology as he points out the difficulties in establishing that these mechanisms are really identifiable even in organisms where interfertility makes genetic analysis possible. But in a sense his treatment is of limited significance because the main problem of homology as a similarity based on ancestry is the matter of the limits of the relationships, which extend far beyond attack by the genetic tests for allelism in genes.

It is our conclusion then that we have no right to claim that we can generally use the term homology in the sense of Lankester's homogeny because the specifications are much too demanding for the kinds of knowledge which are actually obtainable. Whether this has a bearing upon systematics will be discussed in a later chapter. In the meantime, however, after having vigorously attacked the use of the ancestral criterion of homology on the grounds that it is *more* rather than less subjective than the demonstration of essential structural correspond-ence, we must accept many evidences to support the inference of genetic relationship *within* similar structural patterns and use appro-priate terms with which to refer to such inferences. Such terms have already been made available and no new ones need be introduced. For the relationship among individuals with corresponding parts (special homologues) which are of such a nature as to suggest similar genetic mechanisms derived from common ancestry, we suggest mono-phyletic or homophyletic. The use of these terms implies merely that the organisms to which it is applied are believed to have had common ancestry without claiming that they can be traced to a single repre-sentative of the common ancestor or indeed that any common ancestors are now in existence. The term homology is thus the broad one which may include characters which are polyphyletic or monophyletic. It would extend beyond phyla and even larger groups, requiring only as objective a demonstration of structural correspondence as is possible. For example, it would include the "9 + 2" organization of the flagella of a variety of organisms, conclusive evidences for the genetic relation-ship of which will probably never be obtained. By thus limiting the use of the term homology to Owen's meaning, and combining with it the appropriate term to indicate our inferences in regard to ancestry, we can surely be better understood. But an equally great advantage is that greater attention will once more be directed to the nature of the characters themselves. Comparative morphology and systematics will become more scientific and attain their rightful place in biology.

Having made clear that the term homogeny cannot be used in

Lankester's precise meaning, there can still be a general use for it in the sense of common ancestry. That is, it would apply to those characters which appear to have a sufficiently continuous record in time or space so that strong evidences for genetic continuity are available. The meaning would be similar to Haeckel's homophyly, except that it is the characteristics that are referred to rather than individuals or taxa.

The ultimate disregard for Owen's meaning for homology has been reached by Bock (1963) followed by Mayr (1969).

According to Bock, "Contrary to common opinion, the twin concepts of homology and analogy have nothing to do with similarity of features; they are only associated with common origin versus non-common origin."*

For Mayr (1969), "*Homologous* features (or states of the features) in two or more organisms are those that can be traced back to the same feature (or state) in the common ancestor of these organisms."*

Mayr goes on to say that similarity is not a part of the definition of homology because homologous structures are by no means necessarily similar and he cites the case of the ear bones of mammals and the corresponding jaw bones in the lower vertebrates. We are a little puzzled in regard to the adjective "corresponding", if they have no similarity in what do they correspond?

We have criticized Bock already (Boyden, 1969) and will now extend the criticism to Mayr. For the decision to completely eradicate Owen's meaning for homology—as essential structural correspondence—is wrong on three bases, viz.:

1. It is historically wrong. The term homology has implied essential structural correspondence ever since Owen defined the term in 1843! This is (perhaps unwittingly) admitted by Bock (1963) when he wrote, "Contrary to common opinion, the twin concepts of homology and analogy have nothing to do with similarity of features; they are only associated with common origin versus non-common origin." Obviously Bock knew that common opinion said that they were associated with similarity of features. According to Simpson (1961, p. 78), "Homology is resemblance due to inheritance from a common ancestry." Thus

* Permission for the quotations from Bock and Mayr has already been acknowledged.

resemblance has been a part of the meaning of homology for about 130 years and should not be so thoughtlessly discarded. Granted that since Darwin the tendency has been to graft on to homology the idea of common ancestry, and even to define the term with reference to the ancestry, the requirement of similarity was still there.

2. Mayr's definition of homology is etymologically wrong. *Homologous* means same speech—same nature—and such is the correct implication of the term. To propose that the term homologous should have no requirement as to similarity is an unfortunate violation of the proper uses of language.

3. Mayr's definition of homology is wrong on the level of judgment as to how to make the best use of our heritage of words. He is, of course, not alone in this error as the phylogenists generally have been equally guilty. As I have tried to explain (Boyden, 1969) there are two great concepts in post-Darwinian systematic zoology, structural correspondences of many kinds and amounts and genetic relationships from low to high degree. Each of these relationships is worthy of study in its own right and we need appropriate, distinctive and *respectable* terms for each. They are definitely not the same relationships and we do not condone the tendency of evolutionary taxonomists to fail to distinguish between them, or to minimize the importance of structural inter-relationships among the present-day organisms and to exaggerate the inter-relationships which presume to indicate common ancestry.

Now to take homology as the term to indicate common ancestry leaves no satisfactory term for reference to structural correspondences. This was the original meaning of homology and the meaning which justified Huxley's reference to it as "Morphology's central conception". Such terms as "homomorphy" have no commitment to the meaning of essential correspondence, rather they have been commonly used to imply only a superficial correspondence. On the other hand, the central meaning for homogenetic is obvious—it can only refer to similarity in genetic relationships. Sound judgment in the matter of the effective use of our heritage of words should dictate that we use homology and homogeny as their history indicates.

When it comes to analogy the history has been even more confused but Owen did better than most of his successors in using the term to refer to similarities in function or in use of parts. So far as I can determine Owen *never* used analogy as the opposite of homology, as Bock mistakenly claimed, which use has led to endless confusion. Clearly the most unmistakable opposite of homology would be non-homology; structures which do not have essential similarities would be non-homologous. And as to common ancestry—if it is evident, there is homogeny; if not, the characters are heterogenetic. In most discussions, the affirmative uses will be the ones principally employed.

Analogy

A variety of biological meanings for analogy was given in the previous chapter. It is a case of illustrating how scientific language can become completely chaotic; it is as though all was trial and error and no selective agencies were on hand to eliminate the errors. We have commented on these meanings (Boyden, 1943, 1947) as have others, especially Haas and Simpson (1946).

Again it is our chief desire to be understood—that from among the meanings in which analogy has been used we should select one which, if still needed, is least likely to lead to misunderstanding. On this basis we would reject those uses of analogy which attempt to make it the opposite of homology. For opposites, there could be nothing as clear as homologous and non-homologous, to indicate fundamental agreement in structure or the lack of it (Boyden, 1943a). Owen used the term analogy to indicate agreement in function, in the sense of use to the organism, without any implications in regard to structure. It is our belief that the term analogy has much less value than the term homology, and if it is to be used at all, the meaning attached to it should be clearly defined in the reports themselves. But we strongly recommend that analogy never be used as the opposite of homology.

It was Simpson's opinion (1959) that the current usage of homology as defined by common ancestry was acceptable and sufficiently unambiguous. But the continual discussion and re-examinations of its meaning indicate that this is not true. Scholars who approach the concept from a broader point of view tend to support the Owenian usage. Woodger (1929, 1945), Withers (1964) and Jardine (1967) understand homology in terms of structural relationships, especially relative position and connections. It is the ancestry that is inferred from the correspondence, not the reverse, and hence to define homology in terms of ancestry is a rank tautology. Schindewolf (1969) concluded that the concept of homology can only be defined and interpreted on a morphological basis. Inglis (1966) and Cracraft (1967) come to very different conclusions about the usage of homology. The former uses homology as morphological correspondence, the latter violates his own recommendation that "there should be some 'priority' of subject matter included in definitions of a word" and takes the view held by most phylogenists that the concept of homology does not imply any notion of similarity! Cracraft is definitely wrong also in stating that Owen used analogy to indicate a similarity of function in non-homologous structures only. This is absolutely false!

There have been many other suggestions and statements since my report of 1943, but nothing to indicate any better solution to the problem than proposed then, that is, to use homology and analogy as Owen used them. For common ancestry, the terms used by Haeckel, for individuals, or taxa, was *homophyletic;* and for structures, Lankester's term was *homogenetic;* and their adoption without further delay would solve the problem of mutual understanding and be a boon to biologists.

Summary of the meanings of terms
    The results of the objective comparison of the structural characters of organisms may clearly be expressed by the use of the following terms with the meanings stated.
The characters are, in regard to their structures:
    1. Homologous, that is, essentially similar in relative position and connections, and in adult structure and development.
        a. Serially, that is, as linear series along a chief axis of the body of a single individual, or
        b. Specially, that is, as corresponding parts of the bodies of different individuals.
    2. Non-homologous, that is, lacking in the above attributes.
The characters are, in regard to their uses:
    1. Analogous, that is, used for the same purpose, or
    2. Non-analogous, that is, used for different purposes.
The characters are, in regard to their presumed ancestry:
    1. Homogenetic, that is, being of common origin, or
    2. Heterogenetic, that is, of different origin.

With the terms and meanings listed above a morphologist could be understood. On this foundation he may build to any degree of refinement, adding all necessary detail to his descriptions and comparisons. The same would be true of a phylogenist who may have sufficient data for describing his interpretation of the course of evolution as divergent, parallel, or convergent, in particular cases. Moreover, this usage would certainly make it possible to avoid the "endless difficulties" (Hubbs, 1944) in studying close relatives, or confusion which has resulted from the differences in numbers of vertebrae in the backbones of related Vertebrates, or the attempt to use homology in the sense of homogeny, or the conflicts in regard to the embryonic origins of homologous organs. In other words, morphology would achieve its rightful place in zoology as the science of animal structure.

CHAPTER 6

SYSTEMATIC SEROLOGY

Introduction

SEROLOGY is that branch of biology which is concerned with the nature and interactions of antigens and antibodies and antibody-like substances, the lectins. The term is similar in meaning to Immunology but is here used in preference to the latter term which has an implication that immunity is concerned in all reactions between antigens and antibodies. This is so far from the truth that Serology has become the more appropriate term (Boyden, 1948).

Antigens are substances which, under appropriate conditions, induce the formation of antibodies. Most antigens are proteins, but some carbohydrates, e.g., the pneumococcal polysaccharides, are also antigenic. Antibodies, in turn, are proteins, usually produced in response to the presence of foreign antigens in the host organisms, which have the capacity to combine with, and react upon, the antigens used in their formation (homologous reaction) and with chemically related substances (heterologous reactions).* The chief kinds of antibodies now used in systematic serology are gamma globulins, whose synthesis is conditioned by the immunization procedure so that they have new combining capacities or specificities, complementary to the antigens used in their formation. It is thus actually true that we define antigens in terms of their capacity to induce antibody formation, and antibodies in terms of their capacity to combine with and react upon antigens! But these are the distinctive properties of these substances which cannot as yet be readily defined in any other ways.

There are thus only two kinds of primary reagents concerned in systematic serology, and nearly the whole of the science is concerned with the production and testing of these reagents. In addition, however, two

* *Reference* and *Cross-reactions*, respectively, of Williams (1964).

142

other kinds of substances may be concerned in particular tests, viz. complement and haptens. The latter are simple substances which are not antigenic but yet capable of reacting with the antibodies formed to antigens which have chemical similarity to the haptens or which were associated with them during immunization. Complement is a complex of unstable substances necessary for the completion of lytic reactions but of much less importance in the precipitin reaction.

Of the several major kinds of serological reaction (agglutination, anaphylaxis, lysis, complement-fixation, precipitation, toxin–antitoxin) the agglutination and precipitation reactions have been most used in systematic serology, because of their relative simplicity and dependability. The agglutination reaction, titrated by successively diluting an antiserum containing agglutinins until the antigen in the form of suspended particles or cells no longer clumps together, has its most frequent use in the testing and identification of the blood groups of man or other animals and in reactions with bacteria. In these cases no actual titration may be made but only a testing of a constant suspension of cells with the undiluted serum. Corpuscular antigens of great and increasing variety have been described for man (Boyd, 1950, 1962; Race and Sanger, 1968; Stratton and Renton, 1958; Dunsford and Bowley, 1967, and for some other animals, Matoušek, 1965; Tobiska, 1964; 11th European Conference on Animal Blood Groups), all of which seem to be genetically determined. But the systematic relationships are close in these cases, the genetic mechanisms relatively simple, and the systematic applications therefore of limited extent. In the case of bacteria the agglutination reactions may extend beyond the limits of "species" and provide data useful for classification.

It is the precipitin reaction, however, which has thus far contributed most to the development of systematic serology and which has the greatest potential for continued usefulness in the great task of systematizing knowledge about organisms. For this reason, we shall give most of our attention to it. It is a relatively simple, and dynamic, reaction capable of being applied to the comparison of the soluble protein antigens obtained from and representative of any and all organisms.*

* Micro-complement-fixation as used by Sarich and Wilson (1966) is coming on strongly at present.

Discovery and early history

The precipitin reaction was discovered by Rudolf Kraus (1897). He was already familiar with the agglutination reaction, in which formed elements in uniform suspensions such as bacteria or red blood corpuscles, are caused to clump together and settle out following the addition of small amounts of appropriate antisera. It occurred to him to test clear filtrates containing materials extracted from the bacteria, with antisera prepared by the injection of bacterial filtrates into a suitable antibody producer. A very striking result followed; the water-clear mixture of bacterial extract and antiserum began to become turbid very soon and the precipitate thus formed gradually increased and slowly settled to the bottom of the test tubes.

Kraus had several antisera to work with, viz., anticholera, antiplague and antityphoid, and each of these was tested with the antigen used in its formation (the homologous reaction) and with other antigens (heterologous reactions). In each case the antiserum reacted *only* with its own extract. Thus Kraus had produced reagents (antisera) each of which was capable of identifying a particular kind of bacterial extract, in other words, these were *specific* reagents.

Now this is all very well, for an important part of systematics is identification. It may even be that practically the most important part of systematics is identification, for without proper identification a biologist doesn't know what species he is observing or studying and of course therefore he cannot tell anyone else. However, if all serological reactions were thus absolutely specific, kinds of organism could be *distinguished* but not in any way *related* to each other. And most of the rest of systematics does have to do with relationships and with classification.

The absolute specificity of the precipitin reaction was also reported by Tchistovitch (1899), but Bordet (1899) found that an antifowl serum not only reacted with fowl serum but also, though much less strongly, with pigeon serum. This was the first observation of a group reaction and with this slender beginning there has followed seven decades of relationship studies which may have implications for systematics. To Uhlenhuth (1901, 1903) and especially to Nuttall (1904) are we indebted for the more extensive pioneer investigations, but the results of others are not without significance even today.

It is true that in the beginning both Uhlenhuth and Nuttall seemed more interested in identifying antigens than in relating them, but Dr. Graham-Smith, one of Nuttall's early associates, told me this was not true of Nuttall.* At any rate the record clearly shows that, as the precipitin reaction was extended to more species, Nuttall's interest in the systematic application of his data was foremost. All his data as well as the results of studies by his associates were summarized in that classic work, "Blood Immunity and Blood Relationship" (1904). Nuttall's early reports and his book were the source of my own interest in systematic serology. Let me quote just a small section from his report of 1902. In this report he describes the results of the measurement of the volumes of precipitate obtained in homologous and heterologous tests, subsequent to the relatively crude testing of a great variety of animal sera.

"I do not wish these numbers to be taken as final, nevertheless they show the essential correctness of the previous crude results. To obtain a constant it will be necessary to make repeated tests with the bloods of each species and with different antiserums of one kind, making the tests with different dilutions and different proportions of antiserum. I am inclined to believe that with care we shall perhaps be able to 'measure species' by this method, for it appears from the above results that there are measurable differences in the reactions obtained with related bloods, in other words, determinable degrees of blood relationship which we may be able to formulate."

The possibility of thus "measuring species" in a test tube appealed to me and I began fifty years ago to study these possibilities. But here we must have a look at the operational procedures because the data of any science must be judged in terms of their experimental validity. Nuttall himself had suggested the need of further improvements in the techniques of testing, suggestions which he himself never followed.

Progress in methods of precipitin testing

The early tests were very crude indeed. A drop of antiserum was added to a small amount of a saline solution or extract of antigen and

* On the occasion of my visit to Cambridge in 1950.

the developing turbidities were crudely estimated as "Full reaction, marked clouding, medium clouding, faint clouding, or no reaction". (Nuttall, 1904.) The antigen was tested as a 1 : 100 or 1 : 200 dilution of serum (or extract) in saline. There was no knowledge of the original protein concentration of the sera or extracts tested. This has been referred to as a "one-dimensional procedure" (Boyden, 1954b), and it is subject to considerable error. In his "quantitative tests" Nuttall and Strangeways later measured the relative volumes of settled precipitates in capillary tubes with an apparent error of about 10 per cent, but the greater possible errors were not revealed because only one dilution of antigen was tested. As we shall see, comparability of testing would require that antigens are tested in corresponding *relative* amounts and this was not done.

We shall not describe in detail all the variations of technique which have been used in systematic precipitin testing—rather we will mention only representative procedures. Of these, an early modification was that of the ringtest or Schichtprobe first described by Ascoli (1902). Here a series of antigen dilutions is made and a constant amount of antiserum is carefully layered under each. Formation of a layer of precipitate at the junction of antiserum and antigen indicates a positive reaction. The highest dilution of antigen which shows such a layer of precipitate, no matter how thin, indicates the titre of the reaction but not in any quantitative sense its *amount*. However, within the limits of error of the test, the homologous titre is highest, and heterologous titres fall off roughly in proportion to systematic distance. But the antigens tested must be comparable in amounts if the relative titres have systematic validity (Boyden, 1926; Boyden and Noble, 1933; Boyden and De Falco, 1943). Crude flocculation tests may be set up similarly, only in these the tubes containing constant antiserum in increasing dilutions of antigen are mixed. However, in these the results are usually recorded in titres only; i.e., the highest dilutions of antigen which give turbidities or precipitates greater than the controls.

More adequate testing procedures have been developed, viz. the quantitative precipitin technique of Heidelberger and associates and the turbidimetric methods which we have used since 1939. Heidelberger and Kendall (1929) collected, washed, and chemically analysed the precipitates obtained at slight antigen excess or equivalence for given

FIG. 6.1. A photograph of a part of a ring test series. The last tube on the right is the antiserum control tube containing 0·1 ml. antiserum layered under 0·5 ml. saline. Photo by David Trend.

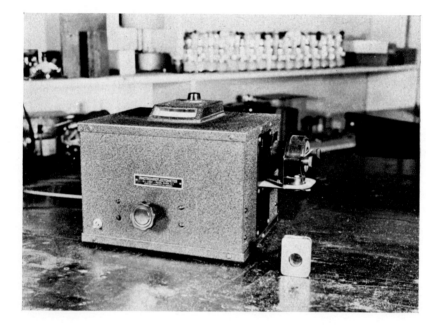

FIG. 6.3. The Libby Photronreflectometer, an improved model. The reagent cells have a capacity of two ml. They fit into the carrier on the right of the photroner, and the turbidity readings are read from the galvanometer on top. The light switch and rheostat knob also show on the front of the instrument.

PROCEDURE

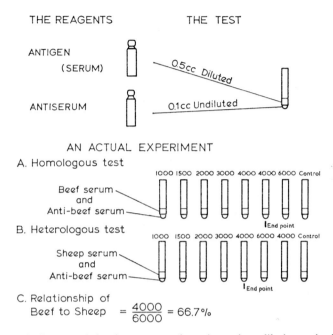

THE REAGENTS   THE TEST

ANTIGEN
(SERUM)

0.5cc Diluted

ANTISERUM

0.1cc Undiluted

AN ACTUAL EXPERIMENT
A. Homologous test

1000 1500 2000 3000 4000 4000 6000 Control

Beef serum
and
Anti-beef serum

End point

B. Heterologous test

1000 1500 2000 3000 4000 6000 4000 Control

Sheep serum
and
Anti-beef serum

End point

C. Relationship of
Beef to Sheep $= \dfrac{4000}{6000} = 66.7\%$

FIG. 6.2. A diagram of the ringtest procedure. An antigen dilution series is prepared with doubling dilutions in each successive tube. The last tube is
the antiserum control, containing 0·1 ml. antiserum layered under
0·5 ml. saline. The titre is the highest dilution of the antigen which
shows a "ring" or layer of precipitate, however fine, at the surface of
the antiserum. The homologous (reference) reaction is the standard
of comparison, to which all heterologous reactions are referred in the
determination of the relationship values. Reproduced from Boyden
(1934, p. 521) with permission of *The American Naturalist* and University
of Chicago Press.

systems. The nitrogen content of the collected precipitate, determined
by microKjeldahl analysis, gives the antibody nitrogen directly if a
nonprotein antigen was used. Where both antigen and antibody are
protein, a substraction of antigen nitrogen must be made to determine
antibody nitrogen. This is, however, still essentially a one-dimensional
technique in its final expression, even though a series of antigen dilu-

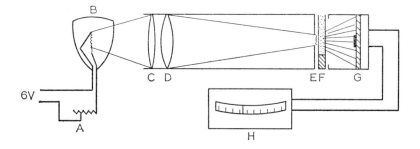

Diagram of the Photronreflectometer

FIG. 6.4. A diagram of the Libby Photronreflectometer slightly modified for increased sensitivity by H. L. Baier.

6 V — Constant voltage transformer supplying a maximum output of 6 volts, maintaining the voltage within ± 1 per cent of the rated value of the transformer. The "photroner" operates between 2 and 4·5 volts.

A — Variable resistor.

B — 32 candlepower automobile headlight bulb.

C — 37 mm diam., 105 mm focal length lens.

D — 53 mm focal length lens.

E — Large central aperture of diaphragm allowing the focused light to strike the black spot whenever the reagent cell is removed, or the particles, when the cell is in place. The small lower aperture allows some light to strike the photocell, when reagent cell is removed for control settings.

F — Reagent cell.

G — Photoelectric cell with black central spot.

H — Galvanometer.

Reproduced from Ph.D. thesis of Dr. H. L. Baier, Rutgers University, 1951. With permission of the author.

tions was used in locating the point at which the analysis was to be made.

In our own turbidimetric procedure (Boyden, 1954b) a really two-dimensional system is used, the turbidity readings being made at a whole series of antigen concentrations covering the entire antigen reaction range. The development of the Libby Photronreflectometer (1938) had provided a sensitive and useful instrument for relative turbidimetric measurements which allowed us to compare entire curves of reaction and not just single points.

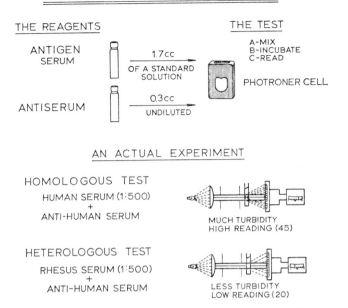

NEW PROCEDURE IN THE SEROLOGICAL STUDY OF ANIMAL RELATIONSHIP

FIG. 6.5. A diagram to illustrate the procedure for measuring turbidity by means of the Libby Photroner. A series of 10 or more doubling dilutions of antigen is prepared and 1·7 ml. of each dilution is placed in the cells. An antiserum control cell receives 1·7 ml. of saline only. The amount of light is standardized to give a selected galvanometer reading (usually 50) with no reagent cell in place after which each cell with its contents is placed in the carrier and the control reading made and recorded. Then into each cell, 0·3 ml. of antiserum is added, and the contents mixed and set aside for incubation. After the stated period, each cell is placed in the photroner and read for its turbidity, after gentle mixing. The net turbidities are determined by subtracting the sum of the antiserum and antigen control values for each dilution. It has been shown that particle sizes are internally compensated and have no effect on the turbidity readings. From Boyden (1942) *Physiol. Zool.* **15**, Fig. 3, p. 113. With permission of University of Chicago Press.

Before we proceed further with these essential matters of technique a brief explanation of the nature of the precipitin reaction would be helpful. The chemical combination of soluble antigen with its homologous soluble antibody occurs in multiple proportions. The quantita-

F

tive relations between the two principal reagents in the precipitates have been studied by Heidelberger and associates and others. The accounts given in Kabat and Mayer (1961) and Boyd (1956) describe these relationships, mainly for homologous systems.

For our purpose, viz. the applications to the study of systematics, it is necessary to understand the following points in regard to comparative precipitin testing. With a constant volume of a precipitating antiserum, the amounts of precipitate increase to a maximum and then decrease to zero following the addition of constant volumes containing increasing amounts of antigen or antigens in saline. There are, in other words, optimal proportions of antigen to antibody which yield maximum amounts of precipitate. Increases of antigen beyond the optimal proportion for any particular system actually decrease the amounts of precipitate formed, and may in fact yield no precipitate at all in the zone of extreme antigen excess! (See Fig. 6.6.)

The current explanation for this behaviour is relatively simple. Precipitating antibody molecules are considered to be bivalent whereas the corresponding antigen molecules are multivalent. If each and every antibody molecule has both of its reactive sites satisfied from the beginning of the addition of the antiserum to the excess antigen solution, the resulting An : Ab : An compound cannot combine with any other An : Ab : An compounds and must remain separate. There is thus no visible result of such combination but the increased molecular size of the combined particles in solution can be shown by their behaviour in the ultracentrifuge, or by their sensitivity to inorganic precipitating agents. Furthermore, analysis of the supernatants in antigen excess reveals no free antibody but, of course, much antigen. But now if the antigen is present in lesser amounts, some of the same antigen molecules will be combined with more than one antibody molecule, perhaps with several, and the macromolecular combinations will aggregate and form visible precipitates. While the rates of original combination are very rapid, the rates of aggregation are slower and are capable of being measured.

There is a definite significance of these facts for comparative serology. Each antiserum must be tested with the homologous and heterologous antigens at comparable relative proportions of antigen and antibody! Such comparable proportions would occur at the region of neutraliza-

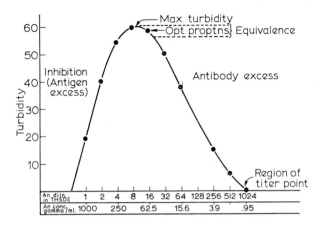

Fig. 6.6. A "normal titration curve" resulting from reacting constant volumes (0·3 ml.) of antiserum with a series of doubling antigen dilutions (1·7 ml.). The entire antigen reaction range extends from zero in the region of great antigen excess through decreasing amounts of reaction in the region of antibody excess. Such complete curves are a guarantee that all antigens present, even in a mixed antigen such as serum, for which there are any antibodies in the antiserum, are present in adequate amounts to completely satisfy the antibody combining capacities. From Boyden (1954) The measurement and significance of serological correspondence among proteins. In W. H. Cole, editor, *Serological Approaches to Studies of Protein Structure and Metabolism*. Rutgers University Press, New Brunswick, N.J. With permission.

tion or equivalence, where in single AnAb systems there is neither detectable antigen nor antibody left in the supernatant. Or the amounts of reaction would be comparable at maximum precipitation (which occurs in slight antigen excess) or in the entire antibody excess zone, or in the entire antigen excess zone, or finally and *best of all*, over the entire antigen reaction range of each antiserum.

With the "normal titration curve" (Boyden, Bolton and Gemeroy, 1947) in mind, let us sketch out the various characteristics of precipitin reaction systems, and the methods for measuring these in saline media. Later we will turn to the precipitin reactions in gelified media and their special applications.

The complete description of any particular homologous precipitin reaction would require that we measure or determine:

The extent or amount of the reaction:
  1. At single points such as (a) titre in extreme antibody excess (b) titre in extreme antigen excess (c) at optimal proportions, neutralization or equivalence, or (d) at maximum precipitation.
  2. At series of points representing substantial portions of the normal titration curve (a) antibody excess region (b) antigen excess region (c) portions of both.
  3. Over the entire antigen reaction range with (a) undiluted antiserum or (b) with successive dilutions of antiserum.

The rate of the reaction:
  1. Combining rates (nearly instantaneous).
  2. Aggregation rates (rapid at first but measurable).
  But how may the "amounts" of the reaction be measured? By one or more of several procedures as follows:

Collection of precipitates formed:
  1. After settling
  2. After centrifuging
  and measured by determining:
  1. Volume
  2. Dried weight
  3. Nitrogen content after Kjeldahl or other chemical analysis.

Turbidities in situ measured by:
  1. Reflection systems, "light scattering", or Libby Photronreflecto-meter
  2. Transmission systems, e.g., B. & L. "Spectronic 20".
  The above characteristics apply to homologous reaction systems whether simple or complex, that is, containing one major kind of antibody only, or containing several kinds of noncorresponding Antigen-Antibody (AnAb) systems. In the latter case a complete description for any particular antiserum would require that the relative amounts and proportional contributions of each kind of system should be determined. Now when heterologous antigens are also tested, as is characteristic in systematic serology, additional properties of the antisera have to be determined, viz.:
  1. Their specificity, i.e., their capacity to discriminate between homologous and heterologous antigens and reactions.

2. Their reaction ranges, viz.:

(a) The ontogenetic reaction range (developmental stages),

(b) The systematic reaction range, i.e., the capacity to react with other species, to the limit of their inclusiveness.

At this point certain additional terms are required before attempting to explain the choice of techniques for application to systematic serology. The more important of these terms are the following:

Serological correspondence. As applied to antigens (or haptens) it exists among all antigens or haptens which react with an antiserum made to any one of the antigens.

As applied to antisera it exists among all antisera capable of reacting with a particular antigen or hapten.

Serological equivalence. This is the highest level of correspondence and is applicable only to systems exhibiting no differences by the procedures used.

Sensitivity. This refers to the capacity of an antiserum to react with minimal amounts of antigen—a capacity which has no simple relationship to the amount of reaction or the content of antibody. Ring test titres are measures of sensitivity.

Avidity. This relates to the rates of reaction (combination and/or aggregation) and has no simple relation to the amount of reactions.

Reciprocal testing. The cross reacting of each of two antisera with the appropriate two homologous and heterologous antigens. For example, antiserum anti-A is tested with antigens A and B; and likewise anti-B is tested with antigens B and A.

With these matters in mind we are prepared to discuss the relative advantages and disadvantages of the various possible techniques used in precipitin testing with special reference to the needs of systematic serology. As we shall see, the evaluation of the results of precipitin testing can only be made in relation to the validity of the methods used.

Choice of techniques for systematic precipitin testing

It is customary nowadays to refer to the Kjeldahl analysis of the nitrogen contents of washed precipitates as "the quantitative precipitin technique" (Kabat and Mayer, 1961) though admittedly the direct weighing of the collected precipitates would also qualify as an absolute quantitative method. As a matter of fact the determination of the nitrogen content of the collected precipitates requires the use of a factor for translation into protein, which in most cases is the average factor 6·25. For AnAb systems containing carbohydrates as antigens or when conjugated with protein, the only correct procedures would involve the use of corrected factors for such antigens, or the method of direct weighing. Similarly for lipo-protein antigens, calculation based on nitrogen content of the precipitates needs significant correction, if it is desired to know the actual amount of the precipitate.

There are in fact a considerable number of reasons for the use of optical methods rather than analytic methods in precipitin testing as applied to systematic serology. The adoption of such methods is not to be construed as denying the essentiality of the analytic methods, for the analysis of the chemical nature of the reaction and the absolute combining capacities of the primary reagents. But after all, relationships, systematic or otherwise, are relative and not absolute, and statistical measures of correspondence or variance are deliberately divorced from their absolute values in order to make them comparable. We list the chief reasons for selecting optical procedures for systematic precipitin testing as follows (Boyden, 1958):

1. All serological correspondence is relative to the homologous reaction as the standard of reference. "Reference" reaction of Williams (1964).

There is no such thing as an absolute measure of serological correspondence or "cross reactivity".

2. There is no direct relationship between the antibody nitrogen content of an antiserum and its specificity or discriminating capacity. Nor, when a variety of precipitating antisera, against different kinds of homologous antigens, are compared, is there any constant relation between milligrams of antibody nitrogen per millilitre and amounts of antigen precipited at equivalence.

3. Any given antiserum will usually be more discriminating among antigens in relative antigen excess than in antibody excess.

Therefore, unless you want a biased value for the measurement of serological correspondence, you must test each antiserum over its entire antigen reaction range (the "normal titration curve").

4. If whole curves are required, the quantitative technique becomes expensive and laborious. At the ends of the curves large amounts of

antiserum are required to provide the minimum amounts of precipitate required for accurate Kjeldahl determination. On the other hand 3 ml. of antiserum (sometimes reduced to 2 or 1 ml.) can provide the entire curve which we use.

5. Turbidimetric readings can begin immediately after mixing, and be continued as long as desired, without the destruction of the system.

To the extent that rates of reaction (aggregation) are important, the measurement of turbidities is the ideal method.

6. By using whole curves and plotting on the geometric series of antigen dilutions, the "areas" under the curve are proportional to the summated turbidities and independent of the absolute antigen concentrations.

(If insufficient antigen is present, the turbidity in antigen excess will not be completely inhibited.)

Finally, the relative amounts of serological correspondence shown among a group of related antigens will depend on the specificity of the particular antiserum used. A consistent relative placement is to be expected, but the values themselves occur on a sliding scale of correspondences. Is it worth so much trouble to use "the quantitative technique" when no two rabbits will be likely to agree?

It seems to us that the whole matter of estimating serological correspondence among antigens is a "set up" for the use of normal titration curves. Furthermore, by its use we can come to the same conclusions in regard to the effects of salt on the reaction, or the study of other properties and effects, as by the quantitative technique (Heidelberger, Kendall and Teorell, 1936; Boyden, Bolton and Gemeroy, 1947).

There is not only a choice of methods for conducting the precipitin reaction in its applications to systematic serology but also in graphically or mathematically expressing the results of testing. The obvious relation

$\dfrac{\text{Heterologous}}{\text{Homologous}} \times 100$ as the figure to be used in expressing relative

amounts of serological correspondence is old and well-established. But on what basis are these relative values to be placed? We have consistently maintained that whole curves should be used in the determination of these relationship values, rather than any single points. The immense advantage of whole curve testing is that such curves are a

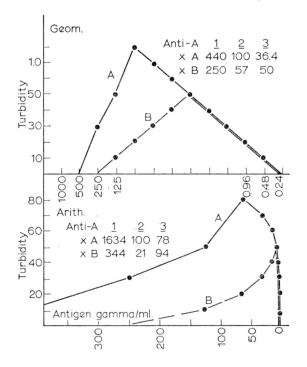

FIG. 6.7. A diagram to show the effects of the methods of plotting the precipitin data on the relative areas and shapes of the resulting curves, and the relationship values for one heterologous reaction B. The top curves represent the usual doubling dilution plot for the relationship of B (heterologous) to A (homologous) antigens. The columns (1, 2, 3) represent the data as follows:

1. Summated turbidities.

2. Per cent relationship of B to A.

3. Per cent of curve area which lies in the antigen excess region.
The lower curves show the same systems plotted on an absolute arithmetic scale. Note the marked effect of the absolute scale on the shapes of the curves and the fact that the antigen excess zones are heavily weighted (78 per cent of area in homologous system and 94 per cent of area in heterologous system). From *Serological Approaches to Studies of Protein Structure and Metabolism*, ed. by W. H. Cole, Fig. 6, p. 83, 1954. Rutgers University Press. Copyright. With permission.

guarantee that every antigen present (even in a mixed system), for which there are antibodies in the antiserum, is known to be present in sufficient amounts to satisfy the combining capacities of these antisera, if the curve descends to zero in antigen excess. Thus it is not even necessary to know the absolute concentrations of the antigens used since they are all adequate for the purpose. Furthermore the method of plotting the graphs need not and should not be on an absolute scale for the information necessary to make such plots correctly is usually not available. For the purpose of plotting on correct absolute scales for the antigens used, the absolute concentration of *each* antigenic component tested must be known, not merely the total antigen concentration. Furthermore the actual relationship values for particular antisera tested with particular homologous and heterologous antigens would be different for the absolute plot, and for the semi-logarithmic plot which we use. This matter can be clarified with reference to a figure from Boyden (1954) which is reproduced here. On the absolute scale the curve of reaction gives relatively great areas to the antigen excess zones and relatively smaller areas to the antibody excess zones. Now since the precipitin reactions are much more discriminating in antigen excess than in antibody excess zones, the automatic effect is to lessen the relationship values in comparison with those revealed for

TABLE 6.1. SUMMARY OF CHARACTERISTICS OF (1) ABSOLUTE ARITHMETIC, AND (2) DOUBLING DILUTION (GEOMETRIC) PLOTS OF REACTIONS MEASURED WITH THE LIBBY PHOTRONER.

(1)	(2)
Curves unsymmetric	Curves nearly symmetric
Long antigen excess zone	Antigen excess and antibody
Short antibody excess zone	excess zones more nearly equal
Area under curve not independent of absolute values of antigen concentration.	Area under curve is independent of absolute values of antigen concentrations if whole curves are used.
Heterologous relationship values relatively less, thus more discriminating among close relatives.	Heterologous relationship values relatively higher, thus less discriminating among close relatives.
Values for serological correspondence are low.	Values for serological correspondence are higher but still representative of the entire reaction.

(Based on whole curves)

From *Serological Approaches to Studies of Protein Structure and Metabolism*, ed. by W. H. Cole, Table I, p. 84. Copyright 1954 by Rutgers University Press. With permission.

the same AnAb systems by the logarithmic plot. Furthermore the latter plot automatically reduces the antigen excess areas and increases the antibody excess areas so that in these "normal titration curves" each of the zones contributes more nearly equally to the final evaluation.

It is appropriate now to refer to and briefly explain the newer techniques in which the reactions are carried on in gelified media and to discuss their values for systematic serology.

Two basic methods for the study of the reactions of precipitin systems in gels were developed by Oudin (1946, 1952) and Ouchterlony (1947, 1958, 1961) and each of these has undergone modifications to adapt them to special purposes. In the Oudin simple diffusion system, the antiserum is mixed with clarified diluted agar, warmed to a temperature allowing solution of the agar and uniform mixing (about 45°C). Capillary tubes which may have been previously lined with a one per cent agar solution and then emptied and dried, are then partially filled with the antiserum agar solution which is then allowed to gel. Such tubes may be stored for weeks in the cold. To begin a reaction the tubes are usually brought to a warmer temperature (room or constant incubator for best results) and the antiserum agar column is overlaid with the antigen solution. If sufficient antigen and antibody are present, one or more zones of precipitate will begin to form at the contact surface and appear to move into the antiserum agar zone away from the interface. In a system containing a single major type of antibody, only a single dense zone may form, with or without separable thin zones representing the minor AnAb systems present. Under proper conditions of relative concentration, the leading zones will represent the more mobile antigens and the following zones the less mobile ones and the separations between them will become greater with time. But it must be realized that under certain conditions zones may coincide, and hence in composite mixtures the number of zones may be only a minimum estimate of the non-corresponding systems present.

For a general presentation of theory and practice the reports of Oudin (1952), Munoz (1954) and Glenn (1957, 1958) and the books of Crowle (1961) and Cawley (1969) may be consulted. Glenn has developed the Serum Agar Measuring Aid (SAMA) by means of which the relative densities of the various zones in agar columns may be measured. A photronreflectometer equipped with a SAMA can be

F‌ɪɢ. 6.8. A simple arrangement for an Ouchterlony plate suitable for testing the similarity of two antigens with an antiserum to one of them. The arc of precipitate lies between the antiserum well below and the two antigen wells above. The reaction shown is a "reaction of identity" indicating the serological equivalence of the two antigens compared. If the antigens had some amount of serological correspondence of a lesser degree, the arc would not be a continuous one but would show minor arcs of different shapes with projecting spurs. From *Serological Approaches to Studies of Protein Structure and Metabolism*, ed. by W. H. Cole. Fig. 8, p. 66. Copyright 1954 by Rutgers University Press. With permission.

operated manually which is, however, a relatively laborious technique. On the other hand the SAMI (Serum Agar Measuring Integrator) moves the column automatically and records the relative densities of all the zones in succession.

The applications to systematic serology are understandable. A particular antiserum is tested with a variety of antigens in different tubes. If this is an antiserum to a native serum, tested with the homologous serum and with heterologous sera, the amounts of reaction expressed in the form of zone densities will decrease in comparison with the homologous standard of reference in proportion to biochemical and apparently systematic distance, the same as in fluid systems. The total densities may thus serve in the same way as the summated turbidities in the fluid systems as the measure of the serological correspondences. Not only may the whole reactions be compared in this way but particular zones may at times also be singled out and compared separately. This can be done quite readily for the albumin anti-albumin zones, which in many mammals lead the way down the columns. To the extent that such comparisons with particular zones are possible, they have the advantage of providing more certain information in regard to relative serological correspondences since they are known to be dealing with comparable systems, and yet do not require the laborious isolation and separate testing of such systems.

The second of the major types of precipitin testing in gels is that of Ouchterlony (1949, 1958). Thin agar plates are used in this procedure and appropriate wells or depots for the antiserum and antigens are provided. A great variety of arrangements of these wells permits special adaptations for particular purposes. In systematic serology it is most useful to use a single antiserum at a time and with it determine the relative amounts of serological correspondence which exist between the homologous antigen and a series of heterologous antigens. For this purpose many arrangements are possible. In one of these, antiserum introduced into a single well is allowed to diffuse through the agar toward the two (or more) antigen-containing wells. The antigen solutions in the meantime, are diffusing through the agar toward the antiserum well. Depending on the relative times of beginning and the rates of these diffusion processes, zones of precipitate will form in the clear agar between the wells, and the number and shapes of the arcs or lines of precipitate provide some of the information desired. Ouchterlony (1961) has recently discussed the types of reaction which result and their interpretation in terms of the immunochemical determinants concerned. Three types of reaction were originally named as follows:

Type I   reaction of "identity"
Type II   reaction of "non-identity"
Type III reaction of "partial identity".

Of these the systematic serologist will be concerned mainly with the Type III reactions, which, if properly performed, can provide information as to the minimum number of AnAb systems reacting and whether these are "identical" or better, serologically equivalent, or not. By a special modification of the wells, Jennings and Kaplan (1961) have been able to obtain a simple relative placement series indicating which of two heterologous antigens is more similar to the homologous. As we shall see later, the major contribution of systematic serology to classification is the providing of such relative placements with the homologous locus as the center of origin. A very extensive study using a comparable technique has been reported by Goodman and Moore (1971).

The resolving power of reactions in gels has been further increased by the combination of (1) electrophoretic separation of a complex antigen in agar or other medium followed by (2) the reaction of antisera with the separated antigens, Grabar and Williams (1953), Williams and Grabar (1955), Williams and Wemyss (1961).

In this procedure a serum or other complex antigen is electrophoretically separated on an agar plate or other medium such as cellulose acetate (Kohn, 1957, 1960, 1961). After the electrophoresis is completed, appropriate antisera are added to troughs in the agar which were provided parallel to the axis of separation of the serum components.

The antibodies present for each antigenic component diffuse toward and meet the antigen zones forming arcs of precipitate at the lateral ends of the antigen zones. Every antigen with a mobility which distinguishes it from the other antigens present will be singled out by corresponding antibodies provided these are present in the antiserum. The conditions may thus be very favourable for the comparison of related sera in that by electrophoresing the homologous and a heterologous antigen side by side followed by the addition of the homologous antiserum the arcs of precipitate indicate the reactions of "partial identity" and also those of "non-identity".* A recent illustration of these methods is available in the report of Williams and Wemyss (1961).

With these representative techniques available for comparison we may return to the matter of the choice of techniques for systematic serology, and with suggestions in regard to the various "levels of approximation" and their uses. Some results of value were obtained more than half a century ago; indeed if this were not so the serological approach would have been discarded long since.

A summary of suggestions in regard to the choice and utilization of serological techniques for the future

The study of related antigens must be carried on in as comparable a way as possible
A. For fluid systems this means, at the very minimum, that tests with each antiserum should be made over the whole homologous and heterologous antigen reaction ranges, or at comparable points in these antigen reaction ranges. In spite of the fact that Nuttall used a one-dimensional procedure, and that it was used by Landsteiner (1936) in his study of the effects of chemical alteration of particular antigens, it is clear that the testing of one and the same arbitrary concentration of the homologous and heterologous antigens can lead to serious errors where these particular concentrations fall on non-corresponding parts of their normal titration curves. If it is decided to test with a single antigen concentration, this particular antigen concentration may be different for each antigen tested

* We prefer more appropriate *serological* terms for the description of these varying amounts of serological correspondence. In our terms the reaction of identity means that the antigens are serologically equivalent since the techniques used were incapable of distinguishing between them. No claim that the antigens were in fact "identical" is made, for in all probability no such identity exists. For the varying amounts of serological correspondence revealed whenever the heterologous antigens react less strongly than the homologous, we would prefer not to use the term "partial identity" for it seems to imply that some of the reacting substances are actually identical and some are not, which may seldom or never be the case. The relative amounts of such serological correspondence can be measured by one or several means as previously indicated.

(whether "pure" or mixed) and must be determined after a trial run of at least part of the normal titration curve.

B. For reactions in gels suitable amounts of reagents should be used in order to get a full reaction.

Thus in the Oudin column, the supernatant antigen solution should be concentrated enough so that *all* zones move away from the interface, but not so concentrated that some of the antibodies are kept in antigen excess and the precipitate kept from forming, or dissolved afterward.

In the Ouchterlony plates when mixed antigens are used, there should be a trial and selection of antigen concentrations which yield maximum density and numbers of zones. The use of a single arbitrarily selected antigen concentration can lead to error here as in fluid systems. This is true even if the antigen concentration selected and used is made following total nitrogen determination of the antigen, for in mixed systems the total nitrogen is no guarantee of equivalence in the separate reacting components, some of which may not even be protein!

The techniques used may be absolute or relative but preferably as simple and economical as possible

In this regard the reactions in gels or cellulose acetate media are the most economical, the optical methods next, and the nitrogen determination least. Thus, a ml. of antiserum or a fraction thereof will suffice for a number of reactions in gels, which may be reduced to a few hundredths of a ml. in tests on cellulose acetate. For the fluid systems a volume of 3 mls. antiserum (which may be diluted if the antisera are powerful) will suffice to cover the entire antigen reaction range of the normal titration curve measured optically, which can be reduced to 0·3 ml. for a single point. On the other hand, for the Kjeldahl analysis of the precipitates formed over the entire antigen reaction range, much more antiserum is needed, especially to define the limits of the curve. For a single point determination, as at neutralization, 3 ml. or more may be needed depending on the strength of the antiserum. Other methods of nitrogen determination than the micro-Kjeldahl may be used to avoid the distillation procedure such as the use of the Folin–Ciocalteu phenol or the biuret reagents.

The tests may be carried out at various "levels of approximation" which provide more and more comparable data.

1. First approximation: the use of mixed antigens (sera, egg white, tissue extracts, etc.) and the antisera made to them.

Such correspondences and relative placement series as are determined in this way can be suggestive only. Their validity can be increased in several ways, each of which establishes greater comparability.

a. The use of reciprocal tests.

A recent example is that described in the preliminary report of Gemeroy and Boyden (1961) in which evidence regarding "the eel problem" was sought. If the reciprocal tests agree within the limits of error it is most likely that both antigens and antisera are comparable. But errors can arise in the determination of relative placement series due to the lack of agreement in the composition of the mixed antigens as explained by Boyden (1959).

b. The selection of sera for testing which show comparability in regard to their electrophoretic patterns.

It is true that the physical separation of the components of sera cannot guarantee their serological correspondence, but the testing of sera each possessing the major proteins as judged by their relative mobilities is preferable to the use of sera without such knowledge (Boyden and Paulsen, 1957). For example, an anti-human serum made against normal human serum would be most likely to possess a relatively large amount of antigamma globulins. Such an antiserum tested with normal serum, and with the serum of an individual devoid of gamma globulins would show considerable differences, quite unrelated to systematic position. It is important in our work to distinguish between the mere presence or absence of particular antigens and the amounts of serological correspondence existing among the antigens present.

c. A further help in the selection of sera for testing would depend on knowledge of the major kinds of antibodies present in the antiserum as revealed by immunoelectrophoresis (Williams, 1956).

With such knowledge the selection of sera for testing can be made on a more intelligent basis. For example, if immunoelectrophoresis reveals no, or negligible amounts of, antialbumins in the antiserum, any test serum can be used regardless of the presence or absence of albumins. The same principle would, of course, apply to any other distinct AnAb system concerned in the tests.

2. Second approximation: the use of partially simplified or "purified" mixed antigens.

The relatively easy separation of albumins and globulins can be used to provide for second approximation studies where the number of independent systems simultaneously reacting has been reduced.

Combined with electrophoresis and immunoelectrophoresis, the greater comparability of results is assured and the systems actually reacting can be identified. But heterologous test sera should be included if they have all the physically separable components which react in the homologous test, even though they do not all react in the heterologous tests! The heterologous protein fractions will indeed lose their serological correspondences at different distances from the homologous species, because their biochemical divergence has occurred at different evolutionary rates. But unless these antigens are present in the distantly related species, their lack of serological correspondence could not thus be established.

3. Third approximation: the use of simplified or relatively purified AnAb systems only.

This would seem to be the final objective of systematic serology, viz., the study of a single AnAb system at a time, and the achievement of relative placements based on such amounts of correspondence as are so revealed. However, words of caution may be necessary here.

a. If the purification procedures denature the antigens even slightly, comparable stages of denaturation will have to be obtained for testing.

b. Every complex antigen molecule is capable of inducing the formation of more than a single type of antibody. In fact, depending upon the immunization procedure used, a whole variety of different types of antibody may result, all of which react with the single homologous antigen, but only some of which react with particular heterologous antigens. Thus the antiserum, unless repeatedly absorbed, will contain a variety of antibodies, with different specificities, all of which are needed to establish the sought-for relative placement series. In fact the systematic reaction ranges extend only in proportion as more kinds of antibodies, and what may be considered a more generalized antiserum, are produced. Of course for identification we want the most marked specificity and discrimination which means the most limited systematic range. But for classification we want more and more extended systematic ranges, to react with, in order of inclusiveness, species of a genus; of a family; of an order; of a class; of a phylum; or even of related phyla.

We see then that the simplification of antisera may extend only to the point of the removal of non-corresponding systems, but that this is not required as long as the reacting systems can be identified. A further objection (besides the additional labour involved) relates to the evaluation of the results for the purposes of classification. The components of a composite system weight themselves automatically whereas if separate systems are used, the problem of giving them a relative evaluation arises. A further discussion of this matter will follow.

Finally, our own procedures, subject to modification, are essentially as follows:
1. To use the fluid systems, with entire antigen reaction ranges, measured optically. (Photronreflectometer, B. & L. spectronic 20.)
2. To use electrophoretic patterns for establishing comparability of antigens and immunoelectrophoresis for the determination of the reacting systems.
3. To search for and attempt to make use of methods, independent of the measurement of serological correspondence, for establishing that the antigens tested are in a comparable physical and chemical state and are truly representative of the organisms from which they were obtained.

Representative results critically examined

We have already intimated that few, if any, of the previous studies in systematic serology meet all present requirements for the demonstration of valid comparative procedures. The task of reporting previous work in detail and evaluating its relative usefulness must be left for a definitive historical summary of the development of knowledge in the field of systematic serology. Here we shall cite representative results only and give a brief commentary on their significance.

To begin with, we refer again to Nuttall, the pioneer systematic serologist. His own studies, together with those of his chief associates, T. S. P. Strangeways and G. S. Graham-Smith, were presented in full in 1904 in the classic, *Blood Immunity and Blood Relationship*. In the final paragraph of his conclusions Nuttall modestly said, "In conclusion I would add that this investigation must necessarily be regarded as preliminary in character. The exhaustive treatment which our present knowledge of the precipitins has received should prove of use to others, and I hope that the work done will stimulate many to further investigate the many problems which present themselves. Like other lines of investigation, this one appeared relatively simple at first; it is evident, however, now, that the phenomena of precipitation are of an exceedingly complex nature."

Nuttall's discussion of the results makes it clear why he considered his findings to be of a preliminary nature: he knew that the concentration of the antigens tested was a source of variation in the amounts and rates of precipitate formation; he knew that in the future "it will be necessary to make repeated tests with the blood of each species and with different antiserums of one kind, making the tests with different dilutions and different proportions of antiserum" (Nuttall, 1902). It remained for others, however, to adopt these recommendations. For samples of his results we refer to the crude tests with five antihuman sera, described in qualitative terms only, and some "quantitative" tests in which the volumes of precipitate were measured after they settled out in capillary tubes. These results were summarized by Nuttall (1904, p. 165).

Among 34 tests with different samples of human serum, 24 gave a "full reaction", 7 gave a "marked clouding" and 3 gave a "medium clouding". All eight tests with Simiidae gave full reactions (orang, chimpanzee and gorilla). Twenty-six species of Cercopithecidae were tested, among which 4 gave a full reaction, 3 gave a marked clouding, 26 gave a medium clouding and 3 were negative. Among nine species of Cebidae, 3 gave marked clouding, 5 gave medium clouding, 2 gave faint clouding and 3 were negative. Among three species of Hapalidae, 1 gave a medium clouding, 1 gave a faint clouding and 1 was negative. Finally, to conclude the tests with Primate antigens, two species of lemurs gave no reaction.

Among the Primate tests then, maximum reactions occurred only with the sera obtained from species in the families Hominidae, Simiidae, and Cercopithecidae, and the *average* intensity of the reactions did fall off in the same way as the systematic relationship based on other characters did. But there was inconsistency, in particular where all eight reactions with Simiidae were full, but not all reactions with human sera were maximum. Again four reactions with Cercopithecidae were full and in these cases obviously the heterologous reactions were greater than those homologous reactions which gave marked or medium cloudings only (10 reactions out of 34). Thus the lack of consistency in the results, as well as the greater reactions of heterologous sera than homologous sera in some cases, suggests a lack of true comparability of the test antigens or the testing procedures, and throws doubt on the general usefulness of the data for systematics. This doubt would be still further increased when it is noted that 14 per cent of the tests of antihuman sera with species of Carnivora, 7 per cent of the tests with Rodentia, and 16 per cent of the tests with Ungulata all gave medium clouding, though the remainder of the tests with these orders gave only faint clouding or no reactions. In these cases non-Primate antigens were giving stronger reactions than some of the Primate antigens did!

From what has been said about variables and procedures in precipitin testing, it should be easy to understand how these inconsistencies could arise. Nuttall used only a one-dimensional test, without in any way being able to place his single test antigen concentration properly with respect to the normal titration curve. This alone could account for certain heterologous reactions exceeding the homologous ones if the former occurred near the peaks of the normal curve and the latter occurred at some distance from the peaks. Furthermore, a variety of procedures was used in the collection and preservation of the antigens—some being relatively fresh and well-preserved fluid sera, others being obtained from filter papers soaked in the blood or serum and then dried.

The crude results were supplemented by quantitative tests (Nuttall, 1904, p. 319) but if the fundamental sources of error previously mentioned were not corrected, it would be expected that inconsistencies would still appear in the results. This is exactly what happened, when the volumes of precipitate obtained from a single arbitrary dilution of the test serum (1 : 40) were compared with the homologous volume as standard of reference. Thus with the reaction of antihuman with human serum as 100 per cent, the reaction with chimpanzee was 130 per cent ("loose precipitum"); gorilla, 64 per cent; orang, 42 per cent, etc. With the non-Primate sera tested the reactions varied from 10 to 2 per cent among species of Carnivora and 10 to 0 per cent among species of Ungulata. In interpreting these results Nuttall wrote as follows (p. 320):

"Among the Primate bloods that of the Chimpanzee gave too high a figure, owing to the precipitum being flocculent and not settling well for some reason which could not be determined. The figure given by the Ourang is somewhat too low, and the difference between *Cynocephalus sphinx* and *Ateles* is not as marked as might have been expected in view of the qualitative tests and the series following. The possibilities of error must be taken into account in judging of these figures, repeated tests should be made to obtain something like a constant. Other bloods than those of Primates give small reactions or no reactions at all. The high figures (10 per cent) obtained with two Carnivore bloods can be explained by the fact that one gave a loose precipitum, and the other was a somewhat concentrated serum."

Other tests with Primate sera were reported by Nuttall but the basic errors in failing to provide for comparability of test antigens were not corrected, and the results must be considered as suggestive rather than as definitive. There were indeed results that were of importance for the future (Boyden, 1951) besides the general and average parallelism between serological results and systematic positions. Thus there was evidence that the whales were more closely related to the even-toed Ungulata than to any other order of Mammalia, a result that we (Boyden and Gemeroy, 1950) have definitely confirmed. There was evidence that Limulus was an Arachnid and not a Decapod; that the serum and egg proteins of birds were somewhat similar to those of Chelonia and Crocodilia. However, for obvious reasons, there is little value in reporting other data obtained by similar procedures for the conclusions to be drawn from them would be no more certain than were Nuttall's. We should pass on then to tests made on a more comparable basis with improved techniques.*

A more comparable basis for testing would be provided if each antiserum is tested with comparable amounts of antigen. An approach to this greater comparability is achieved when the ring test procedure is performed with sera whose total protein content is comparable. This level of testing was reached by ourselves (Boyden, 1926; Boyden and Noble, 1933) in tests with common mammals or with common amphibians. In the first of these reports, ring tests were made with the sera of common mammals, the protein content of these sera having been determined by Kjeldahl determination according to the method of Folin and Wright (1919). Since the variation in the total protein of these sera was within the limits of error of the ring test readings except for one serum, adjustments were required only in this one case.

Samples of results are shown in Figs. 6.9 and 6.10. In Fig. 6.9 two rabbit antisheep sera of different sensitivities are compared, with good agreement, and in Fig. 6.10, an antibeef serum produced in a rabbit is compared with an antibeef serum obtained from a fowl. The results are in fair agreement except for the tests with rat serum. The rabbit antibeef serum gave much less reaction with rat serum than the fowl-produced antibeef serum did. It seems evident that the rabbit-produced reagent was "biassed" against the production of antibodies capable of reacting with rodent sera whereas the fowl was not.

An explanation for these results lies at hand. It is well-known that animals do not readily produce antibodies to their own circulating proteins or to those of closely related animals. Thus the rabbit in responding to injections of beef serum would not be likely to produce precipitins sensitive to the proteins of other species of Lagomorpha, or evidently, even to those of the related order Rodentia. The fowl, however, being so distant from all mammals, would not, when injected with beef serum, be inhibited from producing antibodies capable of reacting with rodent sera, or even with the sera of rabbits! Thus, the

* Ehrhardt (1929, 1930) has summarized the results of most of the reports to the times of his publications. But Ehrhardt treats each report as equal to any other without regard to the relative merits of the testing procedures. We believe a higher level of review is required if the importance of serology in systematics is to be correctly appraised.

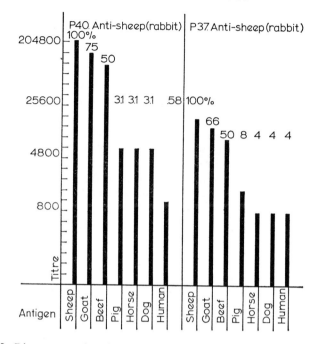

FIG. 6.9. Ring test results of two antisheep sera of very different sensitivities. The antiserum P 40 was approximately ten times as sensitive as P 37. In spite of this the results are very similar. From Boyden (1926) The precipitin reaction in the study of animal relationships. *Biological Bulletin*. Vol. 50, Fig. 6, p. 97. With permission of University of Chicago Press.

fowl would not suffer from the bias or "faulty perspective" (Landsteiner, 1936) that the rabbit would have for the study of mammalian relationships. This point must be borne in mind in systematic serology, for perspective is the one thing that serological placements must have if they are to serve in any important way in systematics.

The second of the reports above (Boyden and Noble, 1933) was concerned with the precipitin testing of the sera of some common Amphibia. Similar procedures were used and the study was directed particularly toward the solution of the problem of the systematic positions of *Siren* and *Necturus*. Noble (1931) was unable to assign a definite place for the families to which these salamanders belong because

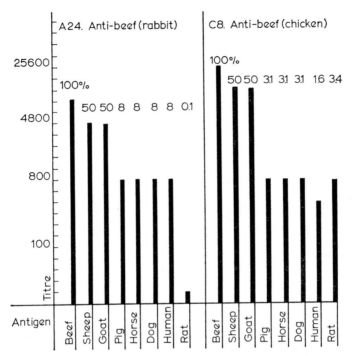

FIG. 6.10. Ring test results with an antibeef serum produced in a rabbit compared with an antibeef serum produced in a chicken. These antisera from such different antibody producers reacted in a comparable way except for the reaction with the rat. From Boyden (1926) The precipitin reaction in the study of animal relationships. *Biological Bulletin*. Vol. 50, Fig. 5, p. 96. With permission of University of Chicago Press.

so many of their structural characters are retained throughout life in a larval form and thus the adult characters are inadequate for proper placement. A new and independent approach to the problem of their relationships was thus required—and found in the precipitin testing of their sera.

In brief, the results indicated that both *Siren* and *Necturus* were far removed from *Cryptobranchus*, an acknowledged primitive form, and belonged rather closer to the more specialized Amphiumidae. Thus an apparently satisfactory solution was found for a problem which was otherwise insoluble and this illustrates how the data of systematic serology may supplement those of systematic morphology and in some cases may be decisive.

Sometime after the study with common Mammalia was completed, an attempt was made to provide a graphic picture of the results in which the per cent relationship values based on the formula $\frac{\text{Heterologous titre X100}}{\text{Homologous titre}}$ were converted into relative distances of the species from each other. The mean values for each pair of species obtained from repeated tests with different antisera of the same kind, together with their reciprocals, provided an average relationship value for the estimate of the distance from the homologous to each heterologous locus. Now since a high per cent relationship value means a close relationship we cannot take the per cent relationship values directly but must take $100 - M$ (M = the average relationship value) as the measure of distance from the homologous locus. On this basis high per cent relationship values would give short distances between the homologous locus and the heterologous loci, and low per cent relationship values would give greater relative distances. It was then discovered (Boyden, 1932) that the relationships among the mammals studied could not be plotted on a plane surface (except by projection) but required three dimensions. The same result was obtained when the results of the amphibian study were plotted. We shall return to this matter in our consideration of the interpretation of serological data and the principles of systematic serology.

The above illustrations indicate that the ring test procedure can give results of interest for systematists but they do not satisfy the requirements for more adequate testing procedures already set forth. The end points in the determination of ring test titres would be the measures of the relative sensitivities of the more sensitive antibodies and they would be valid only if the test antigens were comparable in regard to the concentrations of the particular antigens reacting with these more sensitive antibodies. The total protein contents may be comparable in mammalian sera without there being equivalent concentration of the particular antigens responsible for the titres. Hence errors may arise, reciprocal tests may not agree, and serological placements may not be consistent at such a level of testing. Simplified systems ("purified" or

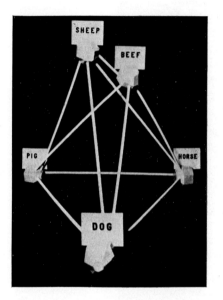

FIG. 6.11. The three-dimensional model resulting from reciprocal ring tests of common mammals. The average 100–M relationship values were as follows:

Beef *vs.* Sheep	31	Sheep *vs.* Dog	95
Beef *vs.* Pig	87	Sheep *vs.* Horse	96
Beef *vs.* Dog	90	Pig *vs.* Dog	94
Beef *vs.* Horse	91	Pig *vs.* Horse	95
Sheep *vs.* Pig	92	Dog *vs.* Horse	95

The model accords with the 100–M values, within an error of ±3 per cent per locus. From Boyden (1934) Precipitins and Phylogeny in Animals. In *American Naturalist*, **68**, p. 524, Fig. 2. With permission of University of Chicago Press.

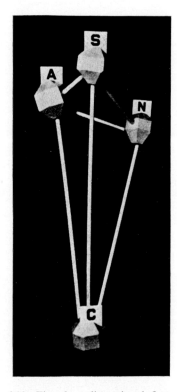
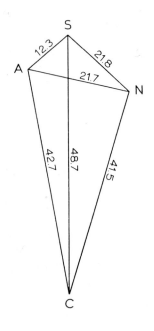

Fig. 6.12. The three-dimensional figure obtained from ring test results with common salamanders. The species tested were as follows:

A. *Amphiuma means.*
C. *Cryptobranchus alleganiensis.*
N. *Necturus maculosus.*
S. *Siren lacertina.*

The 100–M values were halved to give a model of convenient size in centimeters, the values being indicated on the diagram to the left, the model itself on the right. From Boyden and Noble (1933) The relationships of some common Amphibia as determined by serological study. *American Museum Novitates* No. 606. With permission.

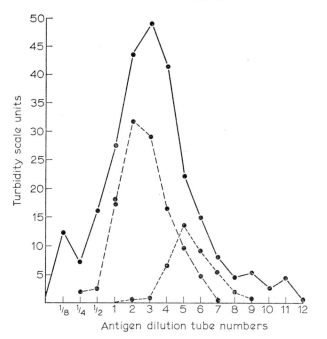

FIG. 6.13. The relationship of *Callinectes sapidus* with *Carcinus maenas* as revealed by photroner testing with an *anti Collinectes* serum. Relative areas were 100, 44 and 14 per cent respectively. From Boyden (1939) Serological study of the relationships of some common Invertebrata. Annual Report of the Tortugas Laboratory, *Carnegie Institution of Washington Year Book*, No. **38**, p. 220.

partially purified) would provide more comparable testing procedures if such antigens can be obtained without significant alteration of their native serological reactivity. But, nevertheless, ring tests give end points in antibody excess only, the procedure is only semiquantitative at best, and practically too much depends on the clarity of the reagents and the skill of the operator in accomplishing a sharp layering of the antiserum under the antigen dilutions. More progress should be possible where a more adequate characterization of the amounts of reaction is possible.

Such progress was made when the Libby Photronreflectometer (Libby, 1938) was developed and used for the study of normal titration curves. Early instances of its use are to be found in the reports of

Boyden (1939, 1942). Figure 6.13 shows the results of testing an anti-blue-crab serum, *Callinectes sapidus*, with the sera of blue-crab, and with *Carcinus maenas*, and *Cancer borealis*. The two members of the Portunidae react most strongly and the representative of the Cancridae is definitely weaker. One striking fact is evident, no *one* absolute antigen concentration can provide an adequate basis for the determination of the relative amounts of reaction which resulted! If the 1 : 500 dilution of the antigens had been selected as the basis for determining the relative amounts of reaction, the relationship of *Carcinus* to *Callinectes* would have been raised to 63% and that of *Cancer* to *Callinectes* would have fallen to zero. If a single antigen concentration was to serve as the basis for the comparison, a comparable point on each curve would have to be chosen, which means practically that a substantial part of the entire reaction ranges would have to be tested in order to make possible the determination of such points. If the peak point had been chosen for these comparisons the relations would have been 64% for the *Callinectes–Carcinus* correspondence and 29% for the *Callinectes–Cancer* correspondence. Neither of these sampling or short-cut methods of determining the relative amounts of serological correspondence does justice to the actual amounts of reaction revealed by the use of whole curves. Impressed as we were with these obvious facts, we thought at one time of paraphrasing a part of a common statement—"the truth, the whole truth, and nothing but the truth" to "the curve, the whole curve, and nothing but the curve". Our view at present is more correct in that we still do recommend the use of curves, and whole curves, but in addition we need more information about the composition of the reacting systems, and the biochemical and biophysical state of the antigens concerned. However, significant results for systematics have been obtained at this level of precipitin testing, results which will be discussed more fully in a later section. We have in mind at this level of testing such studies as are listed in the following table.

TABLE 6.2

REPRESENTATIVE STUDIES BASED ON NORMAL TITRATION CURVES WITH WHOLE
SERA AS TEST ANTIGENS (OR SEED EXTRACTS OF PLANTS)

Authors	Year	Organisms tested
Baier, Joseph G.	1959	Artiodactyla
Baum, Werner	1954	Cucurbitaceae
Boyden, Alan	1942	Horse and Ass
Boyden, Alan	1943b	Crustacea
Boyden, Alan	1953	Pangolin, Aardvark and Armadillo
Boyden, Alan and Douglas Gemeroy	1950	Cetacea and other mammals
Basford, N., J. E. Butler, C. A. Leone and F. J. Rohlf	1968	Coleoptera
Cei, J. M.	1963	Frogs
Cohen, Elias	1955	Reptilia
De Falco, R. J.	1942	Aves
Fairbrothers, D. E.	1966	Nyssaceae
Fairbrothers, D. E. and M. A. Johnson	1961 1964	Grasses
Frair, W.	1969	Turtles
Gemeroy, Douglas	1943	Fish
Gemeroy, Douglas	1952	Primates
Gemeroy, Douglas, Alan Boyden and Ralph De Falco	1955	Primates and Artiodactyla
Gemeroy, Douglas and Alan Boyden	1961	American and European eels
Hammond, David	1955a	Ranunculaceae
Hammond, David	1955b	Solanaceae
Jensen, U.	1968a, b	Ranunculaceae
Jensen, U., D. Frohne and O. Moritz	1964	Rhoeadales and Ranunculaceae
Johnson, M.	1954	Magnoliaceae
Johnson, M. and David Fairbrothers	1961	Cornaceae and Nyassaceae
Kloz, Josef, V. Turkova and E. Klozova	1963	Viciaceae
Kubo, Kazumi	1957	Echinodermata, Sea stars
Kubo, Kazumi	1958	Sea stars
Kubo, Kazumi	1959a, b	Sea stars
Leone, Charles	1949	Crustacea
Leone, Charles	1950a	Crustacea
Leone, Charles	1950b	Crustacea
Leone, Charles	1951	Crustacea
Leone, Charles	1954	Crustacea

continued overleaf

TABLE 6.2 *continued*

REPRESENTATIVE STUDIES BASED ON NORMAL TITRATION CURVES WITH WHOLE
SERA AS TEST ANTIGENS (OR SEED EXTRACTS OF PLANTS)

Authors	Year	Organisms tested
Leone, Charles and Alvin Wiens	1956	Carnivora
Mohagheghpour, N. and Charles Leone	1969	Primates
Moody, Paul	1958	Musk-ox and its relatives
Moody, Paul, Velma Cochran and Helen Drugg	1949	Lagomorpha
Moody, Paul and David Doniger	1956	Porcupines
Moritz, O.	1960	Papaveraceae
O'Rourke, Fergus	1959	Teleostei, *Gadus*
Pauly, L. and H. R. Wolfe	1957	Carnivora
Pauly, L.	1962	Felidae
Shuster, C. N., Jr.	1962	Limulidae
Stallcup, W. B.	1954	Fringillidae
Stallcup, W. B.	1961	Passeres
Wemyss, C. T., Jr.	1953	Marsupialia

The reports listed in the table were mostly based on optical methods of determining the amounts of reaction, the exceptions being the works of Kubo (1958, 1959) wherein determinations of Nitrogen in the precipitates were made. In these cases also, only parts of the antigen reaction ranges were tested, as is the usual custom by workers using these methods. Although without the necessary checks on strict comparability of testing systems, it is probable that most of these results can be used for the determination of systematic placement series, if substantial parts of the normal titration curves were the basis for the comparisons. Recent lists of the literature of systematic serology have been published by Fairbrothers (1969) and Leone (1968).*

Some of the needed knowledge in regard to the composition of the antigens from sera or other sources and in regard to the number of kinds of antibodies available in a particular antiserum has been provided through the development of procedures for the study of precipitin reactions in gels, together with the development of simplified electrophoretic techniques. The latter methods which involve the use of paper or cellulose acetate as the supporting media can reveal the major physically separable components of a serum. The former can provide

* Available for purchase from The Serological Museum, Rutgers University.

a minimum estimate of the number of complementary AnAb systems concerned in the reaction of a particular antiserum with its homologous and heterologous sera. The Ouchterlony plates or Oudin columns, and best of all, the immunoelectrophoretic systems of Grabar and Williams (1953), can provide some of the additional knowledge needed for the determination of the comparability of the testing procedures. A good example of a recent report in which the immunoelectrophoretic procedures were used is that of Williams and Wemyss (1961) who even succeeded in the difficult task of measuring the arcs of precipitate formed at the ends of the antigen zones, as well as in determining the amounts of distinctness or confluence of the lines of precipitate. Our own procedures at present make use of the information provided by electrophoresis and immunoelectrophoresis in regard to the composition of the reacting systems, together with their testing in fluid media by optical methods. Aside from comparability of reacting systems in their composition, there still remains the question of determining, if possible, *independently of serological testing*, the biophysical and biochemical state of the test antigens, in characteristics which do affect the serological reactivity of these antigens.

The types of studies which we need have not as yet been developed to any appreciable extent. These would involve the testing of known systems, viz., "purified" or simplified antigens tested with antisera, the major types of antibodies being identified by immunoelectrophoresis. It is true that we do have quantitative tests with a few purified native antigens such as those of Weigle (1962) but these have been made as partial comparisons, viz., usually in the antibody excess zones only. Since AnAb systems are less discriminating in antibody excess than in antigen excess, this results in higher relationship valves than would be obtained from the same reagents if the normal titration curves were completed. Furthermore in these cases the checks on the number of AnAb systems reacting which could be revealed by immunoelectrophoretic reactions, were not made.

A far better approach to the more adequate comparison of purified antigens is that of Hafleigh and Williams (1966), and of Mohagheghpour and Leone (1969). And to these we must add a series of studies by Sarich and Wilson (1966, 1967a and b) who use purified reagents (albumins) in the study of Primates but with a new technique of microcomplement fixation instead of the precipitin reaction. We will return to these studies later in this chapter.

We have recently had the opportunity to review the matter of specificity in serological reactions and believe the subject must be adequately understood by all those who wish either to use such reactions in their own work or to appraise their actual and potential values for systematics. Accordingly, we present the substance of this report (Boyden, 1971) here.*

The specificity of serological reactions

There are two aspects to the specificity of serological reactions: (a) discriminating capacity by means of which an antiserum is able to distinguish the homologous or "reference reaction" [following Williams 1964] from all others, and (b) the systematic reaction range in terms of which the antiserum is reactive with a few or many related species. Since there have been indications of recent interest especially in regard to the effects of the length of the immunization period on specificity, it may be useful to re-examine the evidences bearing on this matter, and to determine whether my recent statement is valid (Boyden, 1970). "It has been known for more than a half-century that prolonged immunization tends to extend the systematic reaction range of antisera and to lessen their discriminating capacities." Richard Lester (correspondence) has questioned these statements and we will attempt to set forth the principal evidences which support them. In doing so, it will be well to discuss the matter of systematic reaction ranges first; though both aspects of specificity are interrelated.

Systematic reaction ranges

In its first appearances, the precipitin reaction discovered by Rudolf Kraus (1897) was reported as absolutely specific, thus having a minimal systematic reaction range. Tchistovitch (1899) even claimed that anti-horse serum would not react with ass serum! It is difficult to understand how this result was obtained because the amounts of horse serum used

* From Bulletin of the Serological Museum, Rutgers University, New Brunswick, New Jersey, 1971, No. 46. With permission.

in the production of his antiserum were much too large to have produced such a sharply limited specificity. Subsequent workers as ourselves (1953, 1954) most commonly find very high relative correspondences. It might be that Tchistovitch unwittingly used an excess of ass serum in his tests and this inhibited the cross reaction.

So far as we know the first report in which there was any cross reaction was that of Bordet (1899). He had injected a rabbit with chicken serum and discovered that the antichicken serum reacted strongly with chicken but weakly also with pigeon. From such a slight beginning the whole field of systematic serology has developed, though not in the early years for its own sake as much as for the sake of forensic medicine. This being the case the early workers were most often interested in identification and discrimination rather than in systematic relationships. Even Nuttall (1901) seemed to have been first concerned with the medico-legal applications of precipitin testing judging from his first report which concluded, "We have in this test the most delicate means hitherto discovered of detecting and differentiating bloods, and consequently we may hope that it will be put to forensic use". At the time of the above report, Nuttall was apparently unimpressed with the significance of the weak cross reactions which were observed.

But his next reports (1901a, b, 1902) came to different conclusions. He had by 1902 tested 500 bloods with a variety of antisera and stated that "It seems certain that interesting results from the point of view of zoological classification will thus be brought to light". And this work was intensively continued by Nuttall (1904) and his colleagues for several years, culminating in the publication of the classic, *Blood Immunity and Blood Relationship*. In the meantime and during the next two or three decades there was a great deal of controversy in regard to the degree of specificity or aspecificity of the antisera while the technical procedures were becoming more accurate and quantitative. Much understanding of the factors and variables in precipitin testing was obtained however, even with qualitative procedures. We have reported on these matters elsewhere (Boyden, 1942). Uncritical summaries of the reports in the field of systematic serology were published by Erhardt (1930, 1931). The techniques used were various, with both good and bad features and sometimes with conflicting results as should be expected. But in general, the data of precipitin testing gave results in accord with the understanding of systematic relationships in groups which were well studied, thus providing the necessary stimulus to carry further the development of the procedures and their applications to additional groups, plant and animal.

We have at the present time reports of some rather extensive systematic reaction ranges. In animals the reports of Eisenbrandt (1938) and Wilhelmi (1942, 1944) describe certain interphylar reactions, which, though expectedly weak, may be of considerable interest to phylogenists. In one of these, Eisenbrandt reports weak reactions between representatives of the Cestoda and Acanthocephala, and considers that these reactions may indicate a relationship between these groups. The Acanthocephala are placed by Libbie Hyman (1951) in a separate phylum and on this basis, such interactions are interphylar. In any case, no reactions were noted which indicated a relationship of the Acanthocephala with the Nemathelminthes.

Wilhelmi (1940) published a very critical report dealing with helminthes, critical in the sense that he called attention to the fact that Eisenbrandt's test antigens were very low in nitrogen content and his results may have been based on non-protein as well as protein reactions; and critical in the sense that he was careful to use lipoid removal procedures in the preparation of the antigens. His results were consistent in that reciprocal tests generally agreed—which was not the case in the work of Eisenbrandt. But the systematic reaction ranges were relatively narrow, no cross reactions occurring between any of the cestodes and trematodes tested.

It is in the later works of Wilhelmi (1942, 1944) that the greater systematic reaction ranges are reported. In the first of these reports, representatives of three invertebrate phyla, viz., Echinodermata, Annelida and Arthropoda, are compared with representatives of the Chordata. Again lipids were extracted from the frozen dried tissues and saline extracts of such antigens were used in the ring test comparisons. The amounts of antigen actually in solution were determined, and the antisera were induced by means of a single series of four intravenous injections.

The results indicated that the high-titred antisera were capable of cross reactions between phyla and that the Echinodermata were more closely related to the Chordata than were the other invertebrate phyla tested. In the second of the interphylar studies by Wilhelmi (1944), representatives of the Mollusca were compared with those of Echinodermata, Annelida and Arthropoda. Antigens were prepared from tissue extracts as in the previous report. This time a longer series of injections was used in the preparation of the antisera, that is, there were eight intravenous injections with increasing doses. The reason for using a longer series of injections was that the preliminary short series gave no interphylar reactions, and the report of Wolfe (1935) had indicated that a longer series of injections would extend the systematic reaction ranges. We will discuss this matter in some detail further on.

To return to the interphylar reactions, the Mollusca and Annelida were reported to be the more closely related. However, I find it difficult to support this claim from the data of Wilhelmi's Table 1. The fact is that the reaction of the anti-Nereis serum was more than 10 times stronger with *Limulus* than with any mollusc, and reciprocally the anti-*Limulus* serum reacted with Nereis (0·78%) but not at all with *Busycon*! But was it good judgment to select *Limulus* as the sole representative of the great phylum Arthropoda? Ancient as its lineage may be, it can hardly serve as a primitive arthropod. Probably a Crustacean would have been a better choice.

These fragmentary results suggest that further studies with hyperimmune antisera and more quantitative and analytical procedures might provide results of great interest to phylogenists.

Besides such interphylar reactions, several studies have been concerned with interclass reactions. The report by Neuzil and Masseyeff (1958) describes the testing of a hyperimmune horse anti-human serum with sera of representative vertebrates by means of an immunoelectrophoretic procedure. The arcs of precipitate were traced through sera of the four classes, of mammals, birds, reptiles and fish, with most arcs in human serum, viz., 21, and finally down to two arcs in a chicken and two in a lizard and only one in an ostrich and one in a dipnoan. The arcs which carried the farthest were in the albumin zones. More about the relative conservatism of serum albumins versus globulins later in this report.

As to extensive reactions within the subphylum Vertebrata, we may well turn to the lens proteins. It was reported a long time ago by

G

Uhlenbuth (1903) and confirmed by Hektoen (1922) that antilens sera prepared against the soluble antigens of the lens (alpha and beta crystallins) had a very broad systematic reaction range extending throughout all the classes of vertebrates. More recently, Halbert and Manski and their associates have published extensively on this subject. A recent study (Manski *et al.*, 1964) cites the earlier reports and is concerned with the phylogeny of lens proteins. It describes the results of an immunoelectrophoretic analysis of representative lenses (soluble proteins) from all the classes of Vertebrata. The number of arcs in the patterns obtained with an antihuman lens serum declined quite regularly from Primates through other mammals down through a bird (chicken), an amphibian (frog) and two fish (menhaden, carp). Reactions with squid and lobster lenses were negative. The results indicated that the simpler and more mobile alpha crystallines were more conservative than the more complex and less mobile beta and gamma groups. They also showed that the numbers of distinct arcs revealed in the homologous reactions were about the same in all the classes.

Many absorption experiments indicated where characteristic components of one class appeared and how long they remained. For example, all of the different elasmobranch components which persist through mammals and birds were also present in reptiles.

Now between the least and the greatest systematic reaction ranges many intermediate values lie. As we know, it is customary to use the homologous or reference reaction as the standard for comparison and to rate it as 100%. Under comparable conditions, the cross reactions are usually less and indicated as X per cent of the homologous standard. The virtue of such per cent relationships is that they carry the sign of relativity upon them for of course all systematic relationships, serological or otherwise, are relative. The use of logarithms, or any other index of difference *versus* similarity seems to lead some to the belief that such relationship values are absolute. On the contrary, such values belong to sliding scales; for both the relationships and the systematic reaction ranges depend upon the immunization procedures and other variables concerned in comparative serological techniques. This is the next subject for discussion.

The effect of injection procedures on systematic reaction ranges

Perhaps the earliest report indicating that longer immunization periods extend the reaction ranges is that of Magnus (1908). Herein it is reported that an antiserum against *Zea mays* reacted as follows (shown in Magnus, Table 7.3, animal "D").

TABLE 6.3. TWO KINDS OF EVIDENCE BEARING ON ANIMAL RELATIONSHIPS.

Injection period	Positive	Negative
50 days	*Triticum* *Hordeum*	*Phalaris* *Festuca* *Bromus* *Lolium*
109 days	*Phalaris* *Festuca* *Bromus* *Lolium*	*Avena*
130 days	*Avena*	

From Boyden (1963b) Fifty years of systematic serology. Courtesy of *Systematic Zoology* **2**, 19.

The author goes on to state that with even further injections (215 days, 245 days, etc.) the antiserum will react with all *Graminae* . . . but not with any other groups. He commented, "Somewhat more useful is the fact that with ever higher immunization the limits of reaction are ever broadened".

Table I in Magnus' report concerns four different rabbits injected over successively longer periods of time (19, 29, 36 and 146 days) with wheat extract, showing fewer exclusions and more inclusions with the longer injection periods.

Whereas Magnus used crude seed extracts of unknown antigenic composition, a more precise procedure for the comparison of corresponding antigens was used by Lake, Osborne and Wells (1914). Here purified proteins were used as antigens and early and later stages in the immunization of some of the same rabbits were studied. The authors draw certain chief conclusions from their studies of which the following are of greatest interest to us: (p. 374 ff.)

1. Purified preparations of vegetable proteins readily produce antisera. (The proteins used were gliadin of wheat and rye, hordein of barley, edestin of hemp and a globulin of squash seed.)

2. The antisera obtained in their experiments differed in their range of reactions, some giving only the complement fixation, some the complement fixation and precipitin tests, and others in addition conferred passive anaphylaxis to guinea-pigs.

3. Antisera to the same protein obtained from different individual animals differ in their reactions.

4. An antiserum at one stage of its development may be apparently of sharply limited specificity, giving only a reaction with the homologous protein, while a sample taken later from the same animal, after the antibody content has increased, will react with heterologous proteins having similar chemical and physical properties.

Following these conclusions, the authors refer to the work of Magnus cited earlier in this report and view it favourably.

The next report of interest in its bearing on the property of specificity of precipitin reactions is that of Manteuffel and Beger (1922). Using the ring test procedure and the sera of man and common domestic animals as antigens, they prepared a total of 67 antisera, using short injection series (three injections) in some cases and longer injection series in other cases. Though the details of injection for each rabbit are not given, their comments are as follows (p. 371) as we translate:

"Not less important for the prevention of heterologous cloudings is according to our experience the fact that the production of antisera must be quick and continuous. Precipitating antisera which reach the necessary titres in three injections show themselves generally to be more specific than those which require 8 and 10 injections, . . ."

And, further on, they say that the obtaining of frequent heterologous reactions by various workers is due to the great length of the injection period. Indeed one may get high titres this way but the specificity frequently decreases with the length of the injection period. Their whole report is in contradiction to the works of Friedberger and fellow workers who were emphasizing the lack of specificity of precipitin antisera. In the hands of Manteuffel and Beger specificity was the rule as their Table 7 shows. Of the 67 antisera produced, 42 were absolutely specific, 16 gave weak cross reactions and 9 gave moderate cross reactions. Thus they were all absolutely or quantitatively specific.

Hektoen and his fellow workers have contributed much to the study of the specificity of precipitin reactions. We refer to two reports which bear directly on the effects of the length of the injection period. The first of these is that of Boor and Hektoen (1930). A single instance of tests made from an early and a later bleeding of the same animal is presented along with the results of testing several antisera bled at one time only. In essence the results indicated that

dog carbon monoxide hemoglobin antiserum was specific for dog carbon monoxide hemoglobin and dog oxyhemoglobin. However, another sample of blood was drawn from the same rabbit after reinjecting with the dog carbon monoxide hemoglobin. This gave positive tests with human, ox, sheep, rat and horse carbon monoxide hemoglobin, and horse and ox oxyhemoglobin, as well as dog carbon monoxide hemoglobin and oxyhemoglobin.

In this study as well as in the following one by Hektoen and Boor (1931) the ring test procedure was used in testing. Three tables nicely sum up the results of testing a series of antipurified hemoglobin sera, using the hemoglobins of man and monkey and several common domestic animals. Table 1 shows that eight of the antisera gave cross reactions only with the most closely related species, e.g. beef and sheep; horse and mule; man and monkey. Table 2 shows that seven antisera, made in the same way as those reported in Table 1, gave many more cross reactions, even with some distantly-related species. Thus an antibeef hemoglobin serum reacted not only with beef and sheep, but with horse, man, monkey and even with chicken! And an anticat hemoglobin serum reacted with all the species tested! The authors were not able to account for the differences in the specificities of these two sets of antisera and neither can the writer.

However, Table 3 does have a more direct bearing on the property of specificity as affected by manner of injection. This table is entitled "Extension in the range of action of precipitin serum on continued immunization" and it shows that a rabbit injected with one series of injections of beef hemoglobin produced an antiserum which reacted with five of the ten test antigens, whereas the same animal after a second series produced an antiserum which reacted with all.

Continuing with researches in which purified proteins were used, we refer to the report of Hooker and Boyd (1934). Using the ring test procedure for testing a ten-fold antigen dilution series, the authors note that the results agreed with those of Magnus in regard to the effects of continued injections on specificity. We summarize the pertinent results of their study:

Immune sera produced by the injection of 6 times crystallized duck ovalbumin and hen ovalbumin were at first rather specific, those produced by the former apparently absolutely so, as was a serum produced by injecting a hen with duck albumin. Rabbit serum from later bleedings reacted equally with both proteins, but from such sera the reactivity to the heterologous antigen could be removed by specific adsorption, leaving that to the homologous antigen practically unchanged.

Thus even though the use of a ten-fold antigen dilution series in the ring test procedure leaves much to be desired in establishing the amounts of serological correspondence among the antigens, the generalization that a longer series of injections will extend the systematic reaction ranges is clearly supported by the studies of antisera against purified proteins studied separately.

Hooker and Boyd continued these researches (1936, 1941). In the first of these reports the evidences indicate that antisera against purified hen and duck ovalbumins contain multiple antibodies directed against different antigenic determinants. In the second of these reports, the authors give evidences indicating that the later antibodies produced during immunization have a broader reactivity than the earlier ones presumably because the actual combining sites vary more about the main patterns.

During the time that Hooker and Boyd were carrying on these researches with purified reagents, Wolfe was engaged in a long series of studies dealing with the effects of injection methods on the titres, reaction ranges, and specificities of antisera against serum proteins. We refer to one of these reports which bears most directly on matters of range and discrimination (1935). Short series (three or four injections in one series), long series (six or more injections in one or more series) and reinjected rabbits were tested against common carnivores, horse, mule and man. The results indicated that high ring test titres were obtained with rare "extra group" reactions after the first series of injections. On the other hand, reinjection of the rabbits resulted in some loss of specificity. The homologous titres increased but the heterologous values increased much more rapidly.

These are not the only researches which support the general conclusion that the range of reaction of antisera is definitely affected by injection methods. We concur in the conclusion of Harold Wolfe (op. cit., p. 11): that the specificity of a precipitin antiserum depends on the method of injection, and therefore where comparable results are desired similar methods of injection must be employed. Since the study of Leone (1952) bears more directly on discriminating capacities than with systematic reaction ranges, we will postpone discussion of it till further on in our report.

We believe that the studies referred to indicate that the systematic reaction ranges of antisera, whether they react against plant or animal antigens and in the form of purified reagents or composite mixtures, are generally extended by continued immunization. But this conclusion in no way denies that different antigens from the same species may differ in the relative amounts of serological correspondence shown when tested with antisera made in the same way and bled at the same time. In the case of mammalian serum, for example, the evidences indicate that the albumins have been more conservative in evolution than the globulins. In our study of the relative position of the Cetacea among the orders of mammals we (Boyden and Gemeroy, 1950) noted that the antibovine sera which contained the highest proportion of antialbumins gave much stronger reactions with whale sera than did the antisera which contained chiefly antiglobulins. This should hardly cause any surprise in view of the fact that mammalian sera contain perhaps one or two albumins along with dozens of globulins. Obviously more globulins have evolved than have albumins.

Perhaps the most studied of the plasma proteins from the standpoint of their presumed evolutionary history are the Primate albumins. In fact Goodman (1961, 1963, 1964) has developed a very elaborate theory which attempts to explain the relative conservatism of Anthropoid serum albumins on the basis of the dangers of isoimmunization which would presumably eliminate any rapid changes in their constitution and antigenicity. According to Goodman, the human and other higher Primate placentae and the more and more prolonged and intimate relationship of maternal and foetal circulations set up conditions which would tend to slow down the evolution of the serum proteins which appeared early in foetal life, such as albumins, because of the opportunities for immunization of the mother if changes occurred. On the other hand, late maturing proteins, the globulins, which developed after birth, could evolve without any such hindrance and did so according to the theory.

It is an essential premise of Goodman's theory that there actually was a retardation of the evolution of albumins in hominoids in comparison with other Primates. Goodman found no difference in the albumins of man, gorilla, chimpanzee and gibbon and only a slight divergence for the orang-utan. Others have demonstrated such differences, using more sensitive and discriminating procedures than the agar plate methods used by Goodman. Gemeroy (in Boyden, 1958) readily distinguished human serum from chimpanzee and orang sera with two different antihuman albumin sera. These results were apparent from the whole curve areas resulting from photoelectric turbidimetry. We know of no more sensitive and discriminating procedures than these. Hafleigh and Williams (1966) found slight differences between human and pongid albumins (except gibbon) and more marked differences between human and gibbon albumins with a procedure comparable to the use of equivalence reactions only. Mohagheghpour and Leone (1969) reported that human and chimpanzee albumins showed 90 per cent of correspondence when whole curves were used and 97 per cent when optimal proportions only were the basis for the comparison. The greater discrimination of the whole curve procedure is well known to us—we have found in a great many comparisons of related antigens that the antigen excess regions are always more discriminating than equivalence or antibody excess regions. Finally, Sarich and Wilson (1966, 1967a, b) using a micro-complement fixation procedure which is both sensitive and accurate and economical of materials, claim that the albumins of apes, men and various new and old world monkeys have evolved at similar rates and "to much the same degree" in the last 45 million years. They reported slight but significant differences among the hominoid albumins.

In view of these conflicting reports we must have reservations regarding Goodman's theory. Other considerations also suggest that this attitude is best for the present at least. For example we have already referred to the report of Neuzil and Masseyeff (1958) showing that the albumins are the most conservative of the serum proteins down to fish. They were the more conservative of the serum proteins as between Cetacea and Artiodactyla (Boyden and Gemeroy, 1950). The more mobile of the lens proteins were the more conservative. Isn't it possible that the albumin molecules have been more conservative mainly because they are smaller molecules and there are fewer alterations possible within the limitations of their size, charge and physiological functions? Whatever the reason for the regular and conservative evolution of serum albumins, we will continue to make use of their comparison for the more distant relationships.

Specificity in terms of discriminating capacities

To be sure our discussion of systematic reaction ranges has been based on reactions which had some discriminating capacities. But specificity in the latter sense is a complex phenomenon fully worthy of careful and, if necessary, extended consideration.

There are at least two aspects of such discrimination: (a) the determination of the number of distinct AnAb systems and (b) the amounts of correspondence, and by subtraction from the homologous reaction the amount of difference among them. Now the techniques useful in determining the number of systems may not be best suited for measuring the amounts of correspondence between them. For the former the agar gel reactions, in the form of Ouchterlony plates or immunoelectrophoretic procedures, are best. But for the latter some means of measuring the amounts of precipitate formed are needed. With the agar plates the measurements may be made by some densitometric procedures as done by Butler and Leone (1968) and Williams and Wemyss (1961). On the other hand, the use of saline fluid systems lends itself to the collection of precipitates for Kjeldahl analysis or the more direct approach of photoelectric turbidimetry. Sarich and Wilson (1966) use a micro-complement fixation method not involving precipitins.

A great many variable factors are concerned in providing antisera with particular grades of discriminating capacities. We will merely list these first and then proceed with the critical evaluation.

a. The choice of species for study.

b. The choice of antigens from these species.

c. The choice of the antibody producers.
d. The length of the immunization period.
e. The technique of performing the reaction.

Fluid systems
Collection of precipitates for nitrogen analysis.
Turbidimetric comparisons.
Length of incubation period.
Use of titres only as in ring test.
Use of antibody excess region or of equivalence zone or the whole antigen reaction range.

Gel systems
Agar plates and double diffusion or of Oudin columns.

Immunoelectrophoresis
Number of systems (minimum) revealed but densitometric measurements are required for the determination of the amounts of correspondence among the systems.
Reactions of "identity" or "partial identity".

Now to the discussion. There is some choice in regard to the groups and species selected for study. For example in plants one may select a genus of many species, or select genera with few species and study several of them. In the former case one may expect high degrees of correspondence and should select procedures which have relatively high discriminating capacities. In some cases unfortunately, the investigators did not know enough to select such procedures or were unable to use them. This led to wholly unwarranted conclusions in regard to the serological homogeneity of species of the genus *Baptisia*, species which vary considerably in morphology and biochemistry. (Lester, Alston and Turner, 1965.) I have pointed out in a previous report (Boyden, 1970) that the use of a single hyper-immune antiserum (injections made over a 15-month period) could scarcely do anything else but lump them all together—which it did. But the conclusion of "homogeneity" can have no real significance when based on such non-discriminating antisera. In the case of animals similar choices exist. Birds as a class are believed to be more similar, taxon for taxon, than the corresponding systematic categories of mammals or reptiles.

In the former case the use of the more discriminating procedures is indicated.

In vertebrates there is the choice of serum or tissue proteins such as the soluble crystallins of the lens. The latter are conservative and the amounts of correspondence will be relatively great among the classes. Within the Primates and other mammals there is the choice of albumins or some other serum proteins. The former have been the more conservatively evolved and the choice of methods should result in the use of the more discriminating ones. It may be that Goodman made a mistake in selecting the agar block procedures for determining the amounts of correspondence in hominid albumins—differences do exist among them but his "reactions of identity" failed to disclose them.

As to the choice of antibody producers, rabbits and fowls have been the most commonly used and either of these can be induced to produce antisera with useful amounts of discrimination if the proper injection and testing procedures are used. To be sure the fowl would have no rabbit bias and would, from its great distance, be expected to have no mammalian bias of any kind. On the other hand the rabbit might be less stimulated by the antigens of other lagomorphs than fowls would be.

In regard to the effects of the length of the immunization period on discriminating capacities, the study of Leone (1952b) shows that the longer immunization periods tend to cause a lower discriminating capacity. His use of the photronreflectometer in the measurement of turbid systems guaranteed that the testing procedures were adequate to show such effects. With some exceptions, the specificity of antisera generally decreased during the course of immunization. This result is in accord with the great bulk of the evidences previously cited in connection with the discussion of systematic reaction ranges, though measured relatively crudely by ring tests only.

We must now consider the choice of techniques for performing the reactions as they affect the amounts of correspondence or discriminating capacities revealed in the comparisons. We have discussed such topics in previous reports (Boyden, 1954, 1958). The earliest procedures were usually simple mixtures of antiserum and a single dilution of antigen. Later flocculation tests and ring tests were made with constant antiserum and a succession of antigen dilutions. The usual comparison was, however, based on the titre, i.e., the highest dilution of antigen capable of showing a turbidity or a "ring" or layer of precipitate. The sources of error resided in the use of antigens of undetermined concentration as well as in the crudity of the measurements. But even more than this, the titre points lie in extreme antibody excess and this is the least discriminating part of the reacting system.

Later, precipitates were collected and analysed for nitrogen content mostly in the antibody excess region of the antigen reaction ranges after incubation for extended periods of time—24–72 or more hours. (Heidelberger, 1954.) On two counts these comparisons give less discrimination: (1) Antibody excess regions are less discriminating than antigen excess regions; (2) prolonged incubation automatically increases the proportion of heterologous constituents in the precipitates. The early workers who applied the precipitin reaction to the identification of unknown substances soon discovered that the first readings were the most specific and recommended them for forensic medicine. But some workers, seeking for discriminating procedures for the study of close relatives, have not availed themselves of this knowledge.

In 1938, Libby developed the first photronreflectometer at Rutgers University, but it wasn't until ten years later that the Serological Museum was founded and intensive use of it was begun. A simple instrument designed "for the measurement of turbid systems" it proved to have many desirable qualities for the comparison of related antigens. Its performance was critically studied (Boyden *et al.*, 1947; Bolton *et al.*, 1948) and found to be generally satisfactory. Whole antigen reaction ranges were used to establish "normal titration curves", the amounts of reagents required were lowered, readings could be made after very short incubation periods and where whole curves were compared it could be known that all the antigens present were present in adequate amounts to saturate the antibodies if the curves reached zero in extreme antigen excess. For these reasons, we have continued to make the methods of photoelectric turbidimetry our standard for determining the amounts of serological correspondence among related antigens, supplemented now with the methods of diffusion in agar plates or columns and immunoelectrophoresis. But these latter are never considered complete without measures of correspondence for that is what the field of work is all about.

Gel systems usually required prolonged incubation, hence even with densitrometric measurements across the arcs of precipitate (Butler and Leone, 1968; Williams and Wemyss, 1961) they are necessarily less discriminating as to the amounts of correspondence among the systems. That they can determine the minimum number of such systems is granted. But what is the value of knowledge of the number of systems to the taxonomist?

In fact knowledge that, say, four systems were present in a group of species would have no taxonomic importance unless it was demonstrated that they were corresponding systems. But what about the relative placement of the four system species? Knowledge of their correspondence is not enough, the relative amounts of such correspondence must also be known before determining the systematic arrangements.

As we have stated, long continued immunization is likely to induce the formation of greater proportions of heterologous antibodies and this must inevitably reduce the discriminating capacity of the antiserum. To be sure there are other consequences of multiple injections, viz., the less antigenic substances may gradually begin to have a part in the production of additional AnAb systems. Such antisera may become possessed of a higher level of "fidelity" to the complex antigens (Moritz, 1966) and thus tell more about the species being compared. So what is lost in discrimination among the major antigens is in part compensated for by the inclusion of additional systems. And of course the agar procedures can be set up in many ways which are capable of revealing "reactions of identity" or of "partial identity" sometimes with spur formations. As to quantitative measures of the correspondences revealed by such procedures, there is no ready means of obtaining them as yet.

Before we pass on to interpreting the results in the field of systematic serology, we should point out that even when using normal titration curves—now sometimes referred to as the "Boyden procedure"*—the values for serological correspondence depend on the method of plotting the data. Thus we have already pointed out Fig. 6.7 and Table 6.1 that plotting the successive amounts of turbidity with doubling dilutions of antigen, we get an approximation to a normal curve with nearly equal areas in antigen excess and antibody excess. On the other hand,

* So called in Jensen (1968a).

if the antigen concentrations were plotted on an absolute basis much the greater part of the curves would be in antigen excess. This would give maximum discrimination among the antigens but requires knowledge of the absolute antigen concentrations, *all of them*, which the normal titration curves do not.

Interpreting the results

The chemical similarity of the reactive sites of a cross reacting antigen to the reference antigen is the basis for cross reactions just as the degree of complementarity between antigens and antibodies makes the combination possible. But the nature of the similarity is not reavealed in the testing. Since it is at basis a matter of determining *similarities* among related antigens we prefer to use terms which carry this connotation such as serological correspondence or immunological correspondence. Any differences among the antigens compared are secondarily determined by the subtraction of the similarities from the reference values.

It is thus a bit difficult for us to accommodate to the terms "Index of Dissimilarity" or "Immunological Distance" as used by Sarich and Wilson (1966, 1967a, b). But their system for determining the correspondences is quite different from the precipitin procedures. Instead of using constant antiserum tested with the appropriate antigen reaction range, they determine the factor by means of which the concentration of antiserum in any heterologous reaction must be raised to provide a reaction as strong as the homologous standard. The larger this factor, the weaker the heterologous reaction and of course the less similarity of the antigen. Now when it comes to graphs of the spatial relationships of the loci to each other, we come to distances, viz., the 100-mean values for the cross reactions, for obviously the lower these values the greater the distance from the homologous locus. Here we are in much the same situation as Sarich and Wilson reporting mean IDs.

But we must make clear at once that all apparent linear series of cross reaction values obtained with a single antiserum and referred to a single locus may be artefacts! The reason is merely that the reactions based on similarities can be made to determine the amounts but not the natures of, nor the directions of, the differences from the homologous locus. Thus two heterologous loci, each equally distant from the reference locus could be close together or widely separated. Which is the case can only be determined with additional antisera to the loci in question.

Long ago Linossier and Lemoine (1902) noted that the "albuminous substances" of organisms showed a gradual sequence in their distribution and they may have conceived that their actual evolution had been continuous. Indeed it is tempting, when one sees an almost continuous gradation of amounts of correspondence, to interpret it as not only linear but continuous. Take for example this series of values representing the whole curve areas for an antihuman albumin serum tested with a series of Primates (Mohagheghpour and Leone, 1969). 100, 90, 77, 73, 67, 60, 63, 53, 63, 52, 48, 39, 31 beginning with man and ending with *Tupaia*. But this series is neither linear nor continuous any more than are similar series obtained with Artiodactyla or other mammals which have been shown to require three dimensions for the graphical expression of their interrelationships. In other words, the heterologous loci may be in many different directions from the central reference locus.

So what about the gradual evolution of proteins? Where existing organisms represent recently evolved species we might expect to find such gradual changes. Where systematic groups represent aged derivatives only, no such continuities are to be expected. This does not deny that evolution in proteins was gradual when it occurred, but it counsels caution in a too ready assumption that the continuity has been shown.

Whether continuous or not, the comparison of related antigens has given evidences of considerable regularity in their evolution. This conclusion has been reached by Williams and Wemyss (1961), Hafleigh and Williams (1966) and by Sarich and Wilson (1966, 1967a, b). One of the consequences of this behaviour of proteins is that a study of several kinds of antigens from the same organisms will show a parallelism in their relative positions. We noted this phenomenon in a study of several of the plasma proteins of Artiodactyla (Boyden *et al.*, 1950a, 1951), and referred to it as "Parallelism in Serological Correspondence". It expresses itself in the form of a constant sequence of relative placements. We think a better name for the principle would be a principle of *consistency of relative placements*. Of course this has both taxonomic and phylogenetic implications, concerning which we will comment later.

Mohagheghpour and Leone (1969) have confirmed the principle in their study of Primates, and Kloz (1966) for plants.

In a very interesting recent study, Wallace, Maxson and Wilson (1971) report on "Albumin evolution in frogs: a test of the evolutionary clock hypothesis".

In substance the frogs with a relatively limited amount of morphological divergence have as much divergence in their albumins as would be expected if frog albumins had evolved as rapidly as those of mammals! The hypothesis that serum albumins behave as an evolutionary time clock, proposed by Sarich and Wilson (1967a, b), is thus supported. Furthermore there is no essential correlation between the rates of albumin evolution and the organismal evolution shown by the morphology. What should the systematist do with this knowledge? *We* would continue to classify frogs as other organisms on a fundamentally phenetic basis. No single protein, nor even many proteins, however regularly they evolve, can outweigh the phenotype as a whole in taxonomic importance. But they should have their place among the overall character expressions, which total the nature of the organisms.

We may, though perhaps we should not, comment briefly on the application of computering to the data of systematic serology. For example, let us consider the recent study of certain beetles by Basford *et al.* (1968), "Immunological comparisons of selected Coleoptera with analyses of relationships using numerical taxonomic methods". Our first comment is that the adjective "selected" in the title is most appropriate. Not using all the species of any of the taxa studied, there had to be a selection. The result was that, because of the availability of specimens they were testing, only two species belonging to the same genus (*Harpalus pennsylvanicus* and *H. caliginosus*) proved to be related as follows: (a) By immunoelectrophoresis the reciprocal values were 50 and 55·4 per cent; (b) By immunodiffusion the reciprocal values were 78·4 and 70·6 per cent. All the other of the total of 20 species studied were obtained from different genera of different suborders, series, super-families, families, subfamilies and tribes. In other words, there are relatively few closely related species involved in the study. It is no wonder to me then that the series of numbers "resembles a geometric or logarithmic decline in that it is a series of small numbers following a rapid decline". This kind of result was practically guaranteed by the selection of the species tested. We feel that general-izations in regard to the way antisera "see" related antigens might well be held in abeyance until a broader selection of the kinds of taxonomic situations become available. For example, the Primate study previously referred to, Mohagheghpour and Leone (1969), and a study of Crustacean relationships by Leone (1950) gave very different sequences of values.

Further in regard to the beetle study. The immunoelectrophoretic data are relatively crude since only one of two values is given to the densities of the arcs. That is, the reference arcs are all given the arbitrary value of 5 by inspection. The corresponding cross-reacting arcs are given values of 5 or 4, depending upon whether they are visually estimated to be equal to or less than the reference values. Densitometric measurements were limited to the immunodiffusion tests alone. In other words a cross-reacting immunoelectrophoretic arc might be slightly less than the reference arc in which case it would be rated at 4, i.e. 80 per cent of the reference value. But any and all other arcs down to those barely discernible would also be rated as 80 per cent of the reference values! The relationship values for the whole systems revealed by immunoelectrophoresis were the sums of all the arcs; heterologous sum as per cent of reference sum. Is it so surprising then that "the immunodiffusion data correlated better with the morpho-logical data than did the immunoelectrophoretic data". What else could have been expected?

It may be correct to say that no amount of mathematical computering can make up for deficiencies in the original data or for a selection of species which does not represent typical taxonomic situations (if there are any such). To be sure there are many monotypic genera in various animal groups; but on the other hand most animal species are not found in them!

To continue with the interpretation of the data of systematic serology. Basford et al. (1968) speak of a "typical" sequence of per cent relationship values as representing a geometric or logarithmic decline. But this sequence is a consequence of their selection of species from different genera and higher taxa (only one case where two species of a single genus were studied). It has no right to be considered as typical for serological comparisons in general. It is claimed that antisera "see" cross-reacting antigens as such sequences. This is only partly true. It is stated that there is a tendency for antisera to behave "myopically" with respect to their discriminating capacities. But we object to any such opinion. Antisera react in accordance with the kinds of antibodies they contain and with regard to the amounts of chemical similarity in the test antigens with reference to the homologous standard. Neither the nature of the similarities, nor the direction of the heterologous loci from the locus of reference, nor the relationships of various hetero- logous loci to each other are revealed by any single antiserum. Photroner comparisons can be said to measure distances from the homologous locus as proportional to the differences between the reference and cross-reaction values. They are generally more accurate in measuring small to medium amounts of reaction than large ones, for the photroner has a certain range of turbidities within which the galvanometer readings are directly proportional to the amounts of suspended material. The notion that accuracy is less with the weaker reactions or more distant loci is mistaken as far as the photroner is concerned and so is therefore any allegation that antisera are "myopic" in their "seeing". What can truthfully be said is that the photroner has no sense of direction by means of which a single antiserum can locate one or more heterologous loci in space.

The performance of the photroner was well described by Boyden, Bolton and Gemeroy (1947) and Bolton, Leone and Boyden (1948). It was clearly evident in these reports that any discrepancies between the amounts of suspended material and the turbidity readings were greatest at the peaks of the curves, i.e. where the reactions were greatest.

In other reports (Stallcup, 1954, 1961) it has been claimed that ". . . serological studies with the photronreflectometer tend to minimize the differences between distant relatives and to exaggerate the differ- ences between close relatives". We are at a loss to understand the basis for this false conclusion. No data in support of it are presented in Stallcup's reports! Obviously it would require a study of species within a genus to establish that differences between close relatives were exaggerated—there were none of these! At the other end of the scale, the distant relatives should be lumped together if the differences between them were being "minimized" by the photroner; on the contrary

Stallcup shows them widely scattered in all directions! His generaliza-
tions about the photroner cannot be accepted though the data presented can and should be.

We may now be in a position to evaluate the principles of systematic
serology and to this task we now proceed.

Principles of systematic serology

There has been a considerable body of discussion of these (Boyden,
1942, 1954, 1958, 1959; Moritz, 1960, 1964, 1966) and of the interpretation of the data (Goodman, 1961; Williams and Wemyss, 1961;
Jensen, 1968b). We can bring the account up to date by bearing in
mind the recent studies of the relation of DNA and RNA to the production of specific proteins, and by referring to the establishment of
"The Serological Museum" at Rutgers University (Boyden, 1948a and
b, 1949; Boyden, Leone and Bolton, 1949). The Serological Museum is
dedicated to a simple proposition, which is also the most fundamental
of the principles of systematic serology:

1. *The proteins of the bodies of organisms are as characteristic of them
as are any other of their constitutional traits, and they are fully as
worthy of collection, preservation and comparison as, for example, are
the skins and skeletons of those organisms which have them.*

Now the proteins are developed, as recent studies indicate, in
accordance with the DNA code which is in turn carried by the
transfer RNA to the sites of synthesis. In a sense, therefore, these
primordial proteins are the most immediate of all specific character
expressions and they might therefore serve as taxonomic characters
of great significance. Now, the systematic serologist is concerned, not
only with these first produced proteins, but with later generations
of proteins produced by tissue cells and special organs all under the
direction of the DNA code. Even so, the proteins so produced have
been found by the comparative serologist to be systematically distributed among organisms, which provides the fundamental basis for
their value to systematics.

2. *The methods of comparative serology can be used to determine the
relative amounts of biochemical similarity among the proteins or other
antigens of the bodies of organisms.*

We have already detailed these methods. Previous studies have shown, though not always with the desired degrees of accuracy, that serological correspondence is based on biochemical similarities (Landsteiner, 1936, 1962; Wells, 1929; Marrack, 1938) and others.* The relative amounts of serological correspondence give a statistical summation of the particular kinds of biochemical similarities among the antigens, not revealing in the comparison of native protein antigens, how much of the reactivity is due to the chemical composition itself and how much is due to stereochemical arrangements. As has been suggested before, in gross morphology, the relative position of the parts of an organism is the most conservative of the criteria of homology, and this may as well be true for the intimate morphology of antigen molecules.

3. *Serological correspondence in protein antigens is systematically distributed in nature.*

This is another important principle of systematic serology, for even if all the other principles were perfectly true, serological correspondences would have little or no meaning for the systematist, unless these correspondences were found to occur in relation to other kinds of correspondence. That this relationship does exist, as was evident in a suggestive way to Nuttall, is a great help in simplifying the tasks of the systematist, who may therefore be justified in the continuing attempt to put all kinds of character comparisons together in order to attain the most practical and useful classification possible.

The evidences in support of this principle are many, viz., (1) hybrids of known genetic relationship have the serological characteristics which derive from the genetic constitutions of their parents, (2) closely related species as made evident in many of their other characteristics have the more similar protein antigens, (3) serological correspondence does appear to parallel systematic placements where the taxa are well studied and present no particularly difficult problems of relationship.

In all the above statements, protein comparisons have been emphasized, because in most organisms they provide the kinds of information which is most easily related to systematics. Nonprotein antigens or haptens may not be so useful. Thus the group of Forsmann

* See Boyden (1969) for a discussion of some of the limitations of these comparisons.

antigens (Boyd, 1956) is certainly not distributed in accordance with many other biochemical and structural characteristics and a classification based on the distribution of such antigens would be most unsystematic. The blood group antigens, though genetically determined, are often matters of individual expression rather than species criteria, and though useful in population studies have less general significance for systematics.

And in all the above principles, also, we wish to bear in mind that the serological comparisons resulting from precipitin testing should be capable of providing information which distinguishes between the presence or absence of corresponding antigens, and the relative serological similarity of those antigens present. The importance of this distinction is based on simple known facts, viz., (1) that variation in quantities of particular antigens is greater than in their serological correspondences which are in turn based on their biochemical constitutions, (2) that a particular antigen may disappear from among a set of antigens, in particular individuals, without alteration of the specific classification of the organism. Thus in the case of agammaglobulinemia in humans, a whole class of proteins disappears, or practically disappears, in these individuals, while they remain as human beings however unfortunate because their antibody-forming mechanisms are deficient. The studies of electrophoretic patterns in related organisms have revealed instances of systematic significance (Dessauer and Fox, 1954) and others lacking it. The greater systematic usefulness of measuring serological correspondence than of determining mere presence or absence may be based in part on the fact that the absence of a particular antigen creates in that particular organism no internal conflict, whereas the alteration of such an antigen could create such a disturbance as to endanger the life of the organism. Special conditions as discussed by Goodman (1961, 1963, 1964) may have to prevail in order that the organism could thus live and evolve. The recent studies of pathology associated with autoimmunization in humans indicate that these are important considerations in the living of organisms.

Potential contributions of precipitin testing to systematics

We have stated, as well as implied, that the results of precipitin testing must parallel those of other generally used systematic methods if they are to have significance for the classification of organisms. It may be commented, therefore, that we have no self-sufficient data to offer. That if the results accord with existing knowledge they may be accepted, *but* if this is all that precipitin testing can provide, why is it needed? The current opinion of non-serologist taxonomists illustrates this point very nicely; if our results are in agreement with their own interpretations, the serological evidences are gladly accepted. On the contrary, if our serological placements are not in accord with theirs, they may conclude that "seriology" is no good. The fact is that, as is true of systematics in general, the problems are complex and cannot be so easily summarized.

The potential value of systematic precipitin testing varies with the group studied and with the amounts and kinds of information available and possible in regard to each such group. For commercial fisheries, serological information in regard to the number of species or in regard to the size and distribution of interbreeding populations is much sought after, and considered to be important. For systematists generally, interested in the more inclusive categories and in their probable phylogenies, the potential value of precipitin testing tends to increase with the divergence of the organisms. This is especially true where the fossil record is lacking or fragmentary and the systematic gaps are large. In these cases, systematic interpretations and placements are uncertain and conflicting and the results of precipitin testing can tip the balance, so to speak, in support of one particular systematic placement in preference to others. We have had experience in several of these situations. For the interpretation of close relationships a preliminary study of the American and European eels (Gemeroy and Boyden, 1961) is a case in point. Tucker (1959) had recently come to the conclusion that these are a single species, contrary to the views of Schmidt (1922) and most other students of the problem. Our own tests, made reciprocally with Anti-American and Anti-European eel sera, disclosed a considerable amount of distinction between their sera and thus supports the generally accepted view. For the study of more distantly related organisms we have worked with Aardvark, Armadillo and Giant Pangolin (Boyden, 1953b); with the Cetacea (Boyden and

Gemeroy, 1950); the Artiodactyla (Boyden, 1958) or with the Primates (Gemeroy, 1952, 1958).*

The relationship of the whales to other orders of Mammalia has long been a puzzle. The remarks of Simpson (1945) are particularly appropriate and revealing and describe the conditions under which the contribution of precipitin testing to systematics can be of considerable significance. In discussing the systematic position of the Cetacea, Simpson wrote (p. 213):

"Because of their perfected adaptation to a completely aquatic life, with all its attendant conditions of respiration, circulation, dentition, locomotion, etc., the cetaceans are on the whole the most peculiar and aberrant of mammals. Their place in the sequence of cohorts and orders is open to question and is indeed quite impossible to determine in any purely objective way. There is no proper place for them in a *scala naturae* or in the necessarily one-dimensional sequence of a written classification. Because of their strong specialization, they might be placed at the end, but this would remove them far from any possible ancestral or related forms and might be taken to imply that they are the culmination of the Mammalia or the highest mammals instead of merely being the most atypical. A position at the beginning of the eutherian series would be even more misleading. They are, therefore, inserted into this series in a more or less parenthetical sense. They may be imagined as extending into a different dimension from any of the surrounding orders and cohorts."†

We have previously referred to the preliminary results obtained by Nuttall (1904) in which the interordinal relationship of Cetacea to Artiodactyla was relatively high. Nuttall's results as well as our own indicate an unusually high interordinal relationship between the Cetacea and Artiodactyla, which supports one of the older views in regard to the systematic position of the whales and denies the validity of several others. Thus Flower and Lyddeker (1891, p. 233) comment on the lack of evidence to support the assumed Carnivoran relationship

---

* Reactions of antihuman sera tested by Gemeroy are included in the report of Boyden (1958).

† From Simpson, G. G. (1945) "The Principles of Classification and a Classification of Mammals". *Bull. Amer. Mus. Natur. Hist.*, **85**, 213.

of the whales and the positive structural evidences which indicate a relationship to the pig-like ungulates. "Indeed it appears probable that the old popular idea which affixed the name of 'Sea-Hog' to the Porpoise contains a larger element of truth than the speculations of many accomplished zoologists of modern times. . . ." Kellogg's report of 1936 reviews the uncertain and conflicting evidences and tentatively concludes that the Cetacea were descended from some ancient insectivore-creodont ancestry. On the other hand, Mossman's (1937) careful study of the foetal membranes shows the similarity of Cetacea and Artiodactyla and leads him to the conclusion that these orders have descended from a primitive ungulate stock. In our own tests (Boyden and Gemeroy, 1950) it was discovered that the serum albumins are more similar than the globulins of species. This relative conservatism of the albumins has been confirmed in a number of other cases, besides the studies involving the Cetacea.

Among present-day systematic serologists, no one has been more quick to seize upon particular problems of puzzling relationships, though of limited scope, than Paul Moody. The reports listed in our table provide the data which indicate the more probable relationships of the Lagomorpha; the musk-ox; and the new and old world porcupines, and thus contribute to systematic knowledge. The studies mentioned are representative of the potential contributions of precipitin testing to systematics, though most of them were concerned with the testing of whole sera by procedures which do not allow the identification of the major AnAb systems included. When the higher level approaches to precipitin testing are used, we have reason to expect more consistent and comparable results which can lead to consistent placement series.

Systematic precipitin testing does have potential value to systematics and its data can be used to supplement the data of comparative morphology. Its comparisons are obtained from relatively objective measurements, which are, as all relationships are, relative but not absolute. It is here that antisera of the same kind may differ in their specificities, and systematic reaction ranges, thus providing placement series which are consistent in relative positions but not necessarily in numerical relationship values. It is true that each antiserum gives, when tested with a series of related heterologous antigens, what appears to be a linear relationship series, but further testing reveals that this is not

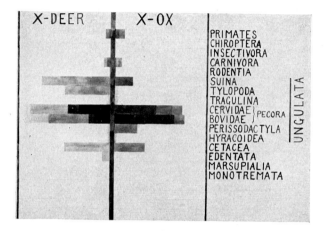

FIG. 6.14. The chart prepared by Nuttall for classroom instruction showing the relative reactions of anti-deer and anti-ox sera tested with representatives of twelve orders of Mammalia. This chart and several others were given to The Serological Museum by Dr. Keilin on the occasion of my visit to Cambridge in 1950. These charts and a number of Nuttall's sera, also given, are preserved as historical documents by The Serological Museum, Bureau of Biological Research, Rutgers University.

a true linear series but instead a multidimensional arrangement. It is true that the testing procedures are limited to existing or relatively recent species and that indeed there is no real measure of primitiveness in serological reactions. But in spite of all these limitations, the systematic testing of protein antigens, conducted with methods which reduce the sources of error to minimum levels, has already been of considerable value in helping to solve difficult problems and may be expected to be of even greater value in the future. It is interesting that systematic serologists have learned to take advantage of the remarkable properties of one great class of proteins, the gamma globulins, in studying the amounts of biochemical correspondence among many other kinds of proteins, the characteristic antigens representative of all kinds of organisms. Complex in detail as the procedures and results may be, they provide additional knowledge in regard to the natures of organisms and can even provide relative placement series for such organisms which may be of considerable value in achieving natural systematic arrangements. What indeed can be a better basis for a natural system than the natures of the organisms themselves? To the extent that systematic serology can contribute to this knowledge, it must be further developed, strengthened and understood.

Current texts in systematics give little attention to systematic serology with one outstanding exception. Crowson's *Classification and Biology* contains a chapter (8) on the Classificatory Use of Non-Structural Characters, and another (13) on Phylogenetic Evidence from Nucleic Acids and Proteins. Crowson believes that the evidences from systematic serology are now more promising than those from DNA hybridization studies, and will continue to be so until the problems of duplication in the codes are solved. We agree with this conclusion. Crowson's discussion is objective and respectable which is more than can be said for the other principle modern systematic texts, for the treatment of systematic serology in the texts of Simpson, Mayr, and Blackwelder are minimal and that in Hennig is completely out of date. References to precipitin testing are all prior to 1938 with a single exception, viz., one sentence from an abstract (that of Boyden, 1959b) which Hennig seems to have misunderstood!

We close our discussion of systematic serology with a table comparing two kinds of evidence which bear upon systematic relationships.

TABLE 6.4—Two Kinds of Evidence Bearing on Animal Relationships

Basis of Comparison	Comparative Serology	Comparative Anatomy
1. Correspondence	Guaranteed in the reactions with protein antigens.	Relation of structural homology difficult to delimit.
2. Objectivity	Relative $\dfrac{\text{heterologous}}{\text{homologous}}$ areas more objective.	Relative similarities and differences less objective.
3. Facility of measurement	Relative curve areas are easily measurable.	Relative morphological indices more difficult to measure and subject to errors of interpretation.
4. Constancy	Relatively constant during the life cycle (checks and balances in antibody-forming mechanisms). Moderately conservative in evolution.	May be greatly varied during the life cycle. Some morphological characters remarkably conservative in evolution.
5. Convergence	Rare.	Common.
6. Relative placement	Apparently consistent.	Not always consistent.
7. Primitiveness versus specialization	No direct test for primitiveness.	Sometimes more primitive conditions may be distinguished from more advanced conditions.
8. Fossil record	Practically nonexistent.	Sometimes well-documented.
9. Requirement for special equipment	Apparatus for quantitative precipitin testing needed.	Sometimes no complicated apparatus needed, but commonly microscopes are required.
10. Testing procedure static or dynamic	Dynamic	Static when descriptive alone. Dynamic when it becomes experimental.

From Boyden (1953b), Fifty years of systematic serology. Courtesy of *Syst. Zool.* **2,** 19.

CHAPTER 7

SYSTEMATICS, EVOLUTION, AND PHYLOGENY

"The more things change the more they remain
the same."   Alphonse Karr 1849.

Introductory

THE THREE terms above are given in the order of their universality. Of the three, only systematics may be said to be universal in biology. Evolution may or may not occur, and the succession of ontogenies in some lines of descent may or may not be determinable with a satisfactory degree of probability, but there must be sound systematics in biology if it is to be accorded the rank of a science rather than that of mythology.

We find support for these statements, first in nature, and second in some of the writings of Simpson (1945).

"It is impossible to speak of the objects of any study, or to think lucidly about them, unless they are named. It is impossible to examine their relationships to each other and their places among the vast incredibly complex phenomena of the universe, in short to treat them scientifically without putting them into some sort of formal arrangement. The science of arranging the myriad forms of life is taxonomy (. . .). A formal classification of animals is a part of taxonomy, but only a part where 'taxonomy' is used in its full sense, and it is also a practical application of some particular set of taxonomic principles, of which many different sets are possible and useful.

"Taxonomy is at the same time the most elementary and the most inclusive part of zoology, most elementary because animals cannot be discussed or treated in a scientific way until some taxonomy has been achieved, and most inclusive because taxonomy in its various guises and branches eventually gathers together, utilizes, summarizes, and implements everything that is known about animals, whether morphological, physiological, psychological, or ecological.

"Classification consists of grouping things according to their characteristics or properties, placing them in a system of categories, and applying a designation to each group thus established" (p. 3).

Blackwelder and Boyden (1952) have suggested a slight change in Simpson's definitions, to avoid using the term taxonomy in broader and narrower senses. They use Systematics in Simpson's broader or full sense for taxonomy.

" . . . systematics is the entire field dealing with kinds of animals, their distinction, classification and evolution".

Simpson (1961) has accepted this change in principle, as indicated by the definitions given in his "Principles of Animal Taxonomy".

"Systematics is the scientific study of the kinds and diversity of organisms and of any and all relationships among them" (p. 7).

He restates the definition of taxonomy (p. 11).

"Taxonomy is the theoretical study of classification, including its bases, principles, procedures and rules."

Now we are in doubt as to the advisability of limiting the meaning of "taxonomy" to the theoretical study of classification, and thus seemingly excluding its actual practise. Mayr's definition (1969) is better balanced; "The theory and practice of classifying organisms". If anything, modern classification has tended to become much too theoretical, because of the current emphasis on its alleged phylogenetic background. The fact is that this emphasis is of long standing. There follows a quotation from a report by W. His (1888) cited with approval in Thompson, *Growth and Form*, 2nd ed., 1951, p. 85.*

"My own attempts to introduce some elementary mechanical or physiological conceptions into embryology have not generally been agreed to by morphologists. To one it seemed ridiculous to speak of the elasticity of the germinal layers; another thought that, by such considerations, we 'put the cart before the horse'; and one more recent author states that we have better things to do in embryology than to discuss tensions of germ layers and similar questions, since

* From *On Growth and Form* by D'Arcy Wentworth Thompson (1951). Published by Cambridge University Press. With permission.

all explanations must of necessity be of a phylogenetic nature. This opposition to the application of the fundamental principles of science to embryological questions would scarcely be intelligible had it not a dogmatic background. No other explanation of living forms is allowed than heredity, and any which is founded on another basis must be rejected."

To be sure the backward look in taxonomy has developed since the publication of Darwin's *Origin of Species* but the astonishing and refreshing fact is that Darwin himself did not spend much energy or time in looking to the ancestry of the barnacles he classified with immense care in his great monographs on the Cirripedia (1851, 1854). Cain (1959b) has remarked (p. 238):

"Darwin as a taxonomist has been very seldom appreciated. He was in fact a good one, and the first part of the thirteenth chapter of the *Origin* (the fourteenth in later editions) is an important document in taxonomic theory." Cain (1959a) had previously discussed this matter in his report on *Deductive and Inductive Methods in Post-Linnean Taxonomy* (p. 208 ff.).

Cain quotes Darwin in part:

" . . . something more is included in our classification than mere resemblance. I believe that something more is included; and that propinquity of descent—the only known cause of the similarity of organic beings—is the bond hidden as it is by various degrees of modification, which is partially revealed to us by our classifications."

Cain comments on Darwin's statement as follows (p. 208):

"In this important passage, an ambiguity lies in the word 'included'. In what sense is propinquity of descent included as something additional in a natural classification? It is obviously not in the classification, but is the *interpretation* of it, and nothing that Darwin says would oppose the idea that this was his view of it. The natural classification was built up by non-evolutionists on natural affinity; the explanation is to be found in the theory of evolution. In the sixth edition this ambiguity has been partly removed; he says that Linnaeus, remark implies 'that some deeper bond is included in our classification than mere resemblance'. I believe that this is the case, and that community of descent—the one known cause of close similarity in organic beings—is the bond, which though observed (?obscured) by various degrees of modification, is partially revealed to us by our classifications!"

Cain has noted (1) that Darwin in the sixth edition qualified his statement about the "only known cause of the similarity of organic beings" by introducing the adjective "close" before the word similarity. This is an indication that Darwin realized his previous statement was untrue for he had discussed at length many kinds of similarities that were not due to "propinquity of descent". And as to this latter term, which had, so far as I can determine, persisted through the first five editions of the *Origin*, in this sentence it is replaced by the term "community of descent", which Darwin had actually used much more frequently in this chapter. These changes indicate to me that Darwin gave serious consideration to these matters in his revision of the sixth edition. That he should replace the term "propinquity of descent" which remained unchanged in the sentence under consideration for five editions, by "community of descent" suggests to me that there must have been some compelling reasons for doing so. Indeed there were, for, contrary to Simpson (1959) and many other evolutionary taxonomists, Darwin never himself classified anything on the basis of "propinquity of descent" only, nor recommended that such be done by anyone else.

We have commented on this matter briefly (Boyden, 1960). We point out that Darwin, in his chapter on "The Mutual Affinities of Organic Beings," used the phrase "propinquity of descent" once and the phrase "community of descent" six times (first edition) and that in the sixth edition the term "propinquity of descent" has been removed and the term "community of descent" is used six times. Yet Simpson quoted only "propinquity of descent" and ignored Darwin's more frequent and mature writing. Even though Darwin was reflecting on the great mysteries of the origin of species for twenty years before he wrote the first edition, he still found it advisable to make such changes in the final edition of his great work.

But let us return to Cain's discussion (1959a) for a moment. Cain has noted the apparent typographical error in regard to the word "observed" in the statement " . . . is the bond, which though observed by various degrees of modification" (sixth edition, p. 205). As Cain suggests (p. 208), the intended word must have been *obscured*. But all this is preliminary to our discussion of Darwin as a taxonomist, for it seems that many biologists have greatly misunderstood what he said about taxonomy and what he actually did about it. Our sources are chiefly two: the monographs on the subclass Cirripedia of 1851 and 1854 and the *Origin of Species*, especially the final edition (sixth).

Darwin as taxonomist

The practise, as revealed in the "Monographs on the Cirripedia"
It is true Darwin had become an evolutionist before he began working on the classification of the barnacles. The two essays of 1842 and 1844 showed that the main features of his evolutionary theory were well in mind before he began his work on the classification of the barnacles in 1846. Even some of the sketchy statements in his notebook of 1837 indicate that the concept of evolution was definitely in mind. Thus, "Propagation explains why modern animals same type as extinct, which is law almost proved. They die, without they change, like golden pippins; it is a *generation of species* like generation of individuals."

"If *species* generate other *species* their race is not utterly cut off" (*Introduction to the Foundations of the Origin of Species*, ed. by Francis Darwin 1909, pp. xi–xii).
To continue with the Essay of 1842, VII (Affinities and Classification):
"Looking now to the affinities of organisms, without relation to their distribution, and taking all fossil and recent, we see the degrees of relationship are of different degrees and arbitrary,—sub-genera,—genera,—sub-families, families, orders, classes and kingdoms. The kind of classification which everyone feels is most correct is called the natural system, but no one can define this (p. 35). . . . Naturalists cannot avoid these terms of relation and affinity though they use them metaphorically. If used in simple earnestness the natural system ought to be a genealogical [one]; and our knowledge of the points which are most easily affected in transmission are those which we least value in considering the natural system, and practically when we find they do vary we regard them of less value" (p. 36).
The much larger Essay of 1844 has its Chapter VII entitled "On the Nature of the Affinities and Classification of Organic Beings," which

clearly indicates that Darwin's whole evolutionary theory was well-outlined as well as his theoretical approach to classification. This chapter concludes:

"Finally, then, we see that all the leading facts in the affinities and classification of organic beings can be explained on the theory of the natural system being simply a genealogical one. The similarity of the principles in classifying domestic varieties and true species, both those living and extinct, is at once explained; the rules followed and difficulties met with being the same. The existence of genera, families, orders, etc., and their mutual relations, naturally ensues from extinction going on at all periods amongst the diverging descendants of a common stock. These terms of affinity, relations, families, adaptive characters, etc., which naturalists cannot avoid using, though metaphorically, cease being so, and are full of plain signification" (pp. 212–213).

These brief quotations from the sketches of 1842 and 1844 clearly show that the outline of Darwin's whole evolutionary theory was clearly in his mind before he began his eight years' work on the classification of the barnacles and that he believed a true classification should be a genealogical one. We might expect therefore to find in his *Monographs on the Sub-class Cirripedia*, repeated claims that particular species are genealogically related to others, and that his groupings are based on knowledge of their "community of descent", or even their "propinquity of descent". Refreshingly, it is mainly in discussing the place of barnacles among other Crustacea that evolutionary theory is obvious.When it came to classifying the species of *Lepas* for example, he gives a straightforward description of the species studied and groups them in accordance with the amounts and kinds of their structural correspondences and their life histories. Perhaps it is significant that on page 3 (Monograph of 1851) he states,

"Although the present volume is strictly systematic, I will, under the general description of the Lepadidae, give a very brief abstract of some of the most interesting points in their internal anatomy, and in the metamorphoses of the whole class, which I hope hereafter to treat, with the necessary illustrations, in detail. I enter on the subject of the metamorphoses the more readily, as by this means alone can the homologies of the different parts be clearly understood."* Darwin concludes his general discussion of the genus *Lepas* with "General Remarks and Affinities"—

"The first five species form a most natural genus; they are often sufficiently difficult to be distinguished, owing to their great variability. The sixth species (*L. fascicularis*) differs to a slight extent in many respects from the other species, and has considerable claims to be generically separated, as has been proposed by Mr. Gray, under the name of Dosima; but as it is identical in structure in all the more essential parts, I have not thought fit to separate it. As far as external characters go, some of the species of Paecilasma have not stronger claims, than has *L. fascicularis*, to be generically separated; and I at first retained them altogether, but in drawing up this generic description, I found scarcely a single observation applicable to both halves of the genus; hence I was led to separate Lepas and Paecilasma. If I had retained these two genera together, I should have had, also, to include the species of Dichelaspis and Oxynaspis; and even Scalpellum would have been separable only by the number of its valves; this would obviously have been highly inconvenient. Although some of the species of Paecilasma so closely resemble externally the species of Lepas yet if we consider their entire structure, we shall find that they are sufficiently distinct; as indirect evidence of this, I may remark that Conchoderma (as defined in this volume), includes two genera of most authors, and yet certainly comes, if judged by its whole organisation, nearer to Lepas than does Paecilasma" (p. 72).

The quotation above from the 1851 monograph shows clearly that Darwin was engaged in making a practical morphological classification and that evolutionary considerations such as matters of "propinquity" or even "community of descent" were kept in the background. Though we know that Darwin had in mind an evolutionary theory of classification, he worked with the species as he found them and classified them in accordance with their observable characters. He speaks of their "more essential parts" (for the purpose of classification that is); of matters of convenience in regard to generic limits, and of placing species nearer or farther from other species as judged by their "whole organisation".

* Homology here is a matter of the correspondence of parts as Owen used the term.

H

This I consider to be a "triumph of the Darwinian method" in a far different sense from that of Ghiselin (1969). But before we enter into that discussion we should add that if any doubt remains that Darwin's Monographs show a practical approach to the classification of the barnacles they need but read any of the descriptions and comparisons of barnacle species as Darwin gave them. Perhaps Darwin realized that, even when he was working with existing species, he had no knowledge of their genealogical relationships and therefore could not classify them in accordance therewith. His deference to facts was clearly shown and this was characteristic of all his attempts to understand evolution, from the very beginning. In support of this statement we turn to his auto-biography for a brief quotation.

"After my return to England it appeared to me that by following the example of Lyell in Geology, and by collecting all facts which bore in any way on the variation of animals and plants under domestica-tion and nature, some light might perhaps be thrown on the whole subject. My first notebook was opened in July 1837. I worked on true Baconian principles, and without any theory collected facts on a wholesale scale, more especially with respect to domesticated productions, by printed enquiries, by conversation with skilful breeders and gardeners, and by extensive reading."

From the same source we quote Darwin's appraisal of his work:

"My work on the Cirripedia possesses, I think considerable value, as besides describing several new and remarkable forms, I made out the homologies of the various parts—I discovered the cementing apparatus, though I blundered dreadfully about the cement glands—and lastly I proved the existence in certain genera of minute males complemental to and parasitic on the hermaphrodites . . . . The Cirripedes form a highly varying and difficult group to class; and my work was of considerable use to me, when I had to discuss in the Origin of Species the principles of a natural classification. Nevertheless, I doubt whether the work was worth the consumption of so much time."

Now let us turn to some of the views of contemporary students of barnacles as to the standing of Darwin's barnacle work. I have re-quested of several living Cirripediologists their appraisal of the current validity of Darwin's monographs. Helpful replies have been received and I wish to quote briefly from this correspondence.

"I have long regarded the monograph as a Superb Systematic Study and apart from the odd detail, as valid today as when written."
"The discovery of new species has merely expanded, not altered, Darwin's views of the group. He laid the foundation on which our knowledge of the group is built."

R. Bassindale
Dept of Zoology
University of Bristol

"It was Charles Darwin who established the morphological nomenclature and systematic concepts upon which subsequent work was based. His two volumes on Recent Cirripedes (1851, 1854) are among the finest morphological and systematic publications in zoological literature. Even after a lapse of more than a century, these works are still among the chief sources of reference."
Quoted from Newman, Zullo and Withers, *Treatise on Invertebrate Paleontology* Vol. R, Arthropoda, pt. 4 No. 1 p R207 (1964), by Victor A. Zullo, who continues (correspondence):
"In essence Darwin took charge of a group that was, at that time, in a state of biological chaos. He presented a clear, logical and meaningful summary (in his usual compendium form) that has essentially defied change. It is an extremely humbling experience to come up with a new idea in a sudden flash of brilliance, only to find subsequently that Darwin had the same thought. He made no truly serious mistakes, keeping in mind species and generic concepts (in the systematic, rather than the genetic sense) of his time. The only fault that I find with Darwin is his ponderous and often ungrammatical prose, which often creates difficulties in ease of understanding of details. This is perhaps the only characteristic that I can claim to share with him."

Victor A. Zullo
Dept. of Geology
California Academy of Sciences
San Francisco

"Darwin's monographs on cirripeds were certainly not 'strictly taxonomic,'* but rather systematic in the finest sense of the word.
* Darwin's term was "strictly systematic".

The works, especially on Recent forms, included a clear attempt at a natural classification including concepts of parallel and convergent evolution, neoteny, and specialization through simplification. There are also extensive discussions on the function and evolution of complemental males. Darwin didn't just classify, he attempted to interpret his findings and explain what he thought had occurred. He was not always right in his interpretations by any means, but the logic he followed after the available facts were assembled leaves little to be desired." W. A. Newman.

Professor Newman then includes the same quotation from the *Treatise on Invertebrate Paleontology* as given in the letter from Dr. Zullo. Thus Newman commended Darwin's inclusion of evolutionary theory, more along the lines of the appraisal of Ghiselin, than my own evaluation which commends the reduction of theory to a minimum in the actual work of the comparisons and groupings.

In this connection it is of interest to note Darwin's own opinion as to the importance of evolutionary theory in his barnacle monographs. We quote:

"As to our belief in the origin of species making any difference in descriptive work, I am sure it is incorrect, for I did all my barnacle work under this point of view. Only I often groaned that I was not allowed simply to decide whether a difference was sufficient to deserve a name".

> *More Letters of Charles Darwin*
> Vol. I, p. 453. Ed. by Francis
> Darwin (1903).

The quotation above is from a letter to J. D. Hooker dated March 30, 1859. I cannot be absolutely certain about its interpretation and my efforts to support my opinion from Hooker's own letters have not been successful. I believe Darwin was saying that his evolutionary theory made no difference in his descriptive work with the barnacles—or at least, the effects of the theory regarding the Origin of Species, much in mind while actively writing this great work, were minimal. But the second sentence about his "groaning" does imply that he could not work on a purely descriptive basis. From my point of view the barnacle monographs are excellent examples of systematic works that are mainly descriptive with evolutionary theory kept to a minimum level. We shall return to the discussion of Ghiselin's views later. Here let us resume one more instance of further quotations from our correspondence with cirripediologists, in support of the excellence of Darwin's monographs. To be sure this is hardly necessary.

"There is no doubt in my mind that Darwin established and rationalized the systematics of cirripedes on the firmest of foundations. Previous to his undertaking the revision of the group, the systematics of barnacles was in a very poor state. The systems of classification as devised by previous workers did little to further any understanding of where cirripedes should fit in the hierarchy of organized beings . . ."

"Apart from the fact that many more species have been added to Darwin's list and more intensive specialized work has been done on various aspects, the monographs basically still serve as the 'prime' source of reference to cirripediologists. Essentially the systematics are still valid and closely followed by contemporary workers. Where major taxonomic changes have taken place are in those very groups in which Darwin encountered particular problems. In the main, however, the basic principles of his systematics remain as important and useful today as ever it did. I doubt if there is a paper written today on the taxonomy of cirripedes in which the author does not have recourse to quote some comparative fact as first pointed out by Darwin. So, with complete justification, I believe I can honestly say that the Monographs will stand for a very long time to come, as a trustworthy monument to Darwin's systematic acumen."

Richard Read
43 Holly Terrace
Hensingham, Whitehaven
Cumberland, England

These quotations from present-day students of the barnacles provide ample evidence that Darwin's monographs on the sub-class Cirripedia have met the test of time and have served as systematic works of solid value. I have been most interested in them because the foremost evolutionist of his day, fully aware of much of the theory of evolutionary classification, yet confined himself largely to the "present nature" of the barnacles he studied, grouping them in accordance with their demonstrable characteristics, and reducing theoretical considerations in regard to their ancestry to a minimum. In my opinion this is most commendable and my regret is that more of Darwin's successors have not followed his example. Nowhere does he claim that his classification *shows* the "genealogical" relationships of the species classified; nor could he prove this even if it were true. What then of the relation between Darwin's practise of classifying and his evolutionary theory of classification as set forth in *Origin of Species?* Is there really such a fundamental contrast between them as appears at first sight? We must discuss this topic with some care, but first we must turn to Ghiselin's discussion of Darwin's method and especially to his chapter on barnacles.

Ghiselin (1969) is convinced that Darwin's "entire scientific accomplishment must be attributed not to the collection of facts, but to the development of theory". "Unless one understands this—that Darwin applied, rigorously and consistently, the modern hypothetico-deductive scientific method—his accomplishments cannot be appreciated."

My own evaluation is somewhat different. In the first place Darwin's time during the voyage of the beagle was largely taken up by observation and collecting facts. Hypotheses were in mind or developing but these were being tested against facts. How else can one test theory? To continually praise the theory and belittle the facts as Ghiselin does is wholly unwarranted. The eight years' work on the barnacles was also largely concerned with an exhausting study and comparison of the barnacles in hand and their grouping in accordance with the facts of their "whole organization".

In the chapter on Barnacles, Ghiselin states (p. 105) "Nowhere in his writings on cirripedes does Darwin discuss, in explicit terms, the relationships between evolutionary theory and the biology or systematics of barnacles. His references to such matters are there, but covert, scattered, and cast in the form of an older terminology, for these works were published before the *Origin of Species.* Nonetheless, one may, by careful reading, and especially by analogy with statements occurring elsewhere, reconstruct his ideas on comparative anatomical method and the evolution of barnacles and demonstrate their connections with the rest of his thought."

My comment here is that the fact that the barnacle monographs were published before the *Origin of Species* probably has nothing to do with the omission of evolutionary theory from them. Darwin's evolutionary theory was quite completely worked out in his *Sketch* of 1844. If he had wanted to insert an evolutionary theory of classification into the barnacle monographs he could have done so. In my opinion the elimination was deliberate as suggested by that comment in his letter to J. D. Hooker previously quoted. Let us repeat it as it is well worthy of remembering.

"As to our belief in the origin of species making any difference in descriptive work, I am sure it is incorrect, for I did all my barnacle work under this point of view." (*More Letters of Charles Darwin*, Vol. I, p. 453.)

Ghiselin agrees with the views of the cirripediologists previously quoted in saying, " . . . his monograph has long remained one of the standards of excellence in taxonomy. Its interest is more than historical, for it is still an indispensable reference, and it continues to provide useful ideas for those who are carrying on the study of barnacles" p. 104.

Ghiselin continues with his commentary on the barnacle works by calling attention to Darwin's figures 4 to 8 inclusive (1854, p. 39).

Ghiselin states:

"The diagrams, adapted from the work of one of his predecessors, are a traditional representation of what he called a 'homological plan', depicting only certain features common to the members of a group and not necessarily asserting anything about historical origins. Each basic part in any form can be seen to correspond to a similar part (such as a whole plate or an articulation) in some other species, although some parts may be absent or fused. The corresponding parts (technically, 'homologues') are labelled with the same letter. What the depicted correspondence means has long been a source of confusion in the philosophy of biology. To a Platonist the plates can all be related to an ideal organism, and the relation between them is one of 'essential similarity'. To Darwin, and to the modern evolutionary systematist, such an interpretation is nonsense."*

My comment here is that Ghiselin has gone too far in claiming that

* All quotations from *The Triumph of the Darwinian Method*, copyright University of California Press. With permission.

morphological correspondences indicating homological plans, types, archetypes and essential similarities were "nonsense" to Darwin. He used all these terms properly and without apology. The following statements by Darwin will support me.

"But it still remains undecided what rank in this class (Crustacea) Cirripedes should hold. Before briefly discussing this point, it is indispensable to indicate their essential characters which I will immediately attempt" (1854, p. 10).

"In the order Thoracica, the abdomen is quite rudimentary, though often still bearing caudal appendages; in the pupa, however, of this order, as in the mature animal of the two other orders, it is formed of three segments. Hence I conclude that, notwithstanding the absence of the above two segments with their appendages in the Thoracica, the archetype Cirripede may be safely said to be composed of seventeen segments" (1854, p. 11).

"In two out of the three orders into which Cirripedes may be divided, the mouth is succeeded, in the adult animal, by eleven most distinct segments; of which the first (i.e., the seventh cephalic) differs from the seven succeeding thoracic segments; and these seven again differ from the three abdominal and terminal segments. Hence it must be admitted that, as far as the cephalo-thorax of the archetype Cirripede is concerned, it consists, like that of the archetype Crustacean, of fourteen segments, of which eight succeed the first-named six that form the mouth and front of the head; and that with the three abdominal segments, there are altogether seventeen segments" (1854, p. 111).

"*Platylepas decorata.*—This genus is closely allied to Coronula and the cementing apparatus is essentially similar" (1854, p. 143).

"I curiously examined the cementing apparatus in *Balanus galeatus, improvisus* and *cranatus*, which have all calcareous bases, and belong to different sections of the genus; and the structure seemed to be essentially the same" (1854, pp. 150–151).

Now it should hardly be necessary to multiply these quotes to prove that Darwin was using the appropriate terms for systematic study in meaningful ways. Let no one claim that when speaking of an "archetype Crustacean" Darwin had some metaphysical concept in mind. His thoughts and acts were very largely concerned with comparisons of the organisms as he found them and when he became phylogenetic in his thinking he was inclined to say so. Thus,

"I hope to be excused for describing at such length, the apparatus by which sessile cirripedes are permanently attached to a supporting surface; for this is the great leading character of the sub-class, not hitherto observed in any other Crustacean. It is not easy to overstate the singularity and complexity of the appearance of the basal membrane of a Balanus or Coronula; and when we consider the homological nature of the apparatus, the subject becomes still more curious: I feel an entire conviction, from what I have repeatedly seen in several genera of the Lepadidae, both in their mature and pupal condition, and from what I have repeatedly seen in Proteolepas, that the cement glands and ducts are continuous with and actually a part of an ovarian tube, in a modified condition; and that of the cellular matter which, on one part, goes to the formation of ova or new beings, in the other and modified part, goes to the formation of the cementing tissue. To conclude with an hypothesis—those naturalists who believe that all gaps in the chain of nature would be filled up, if the structure of every extinct and existing creature were known, will readily admit that Cirripedes were once separated by scarcely sensible intervals from some other, now unknown, Crustacean. Should these intervening forms ever be discovered, I imagine they would prove to be Crustaceans, of not very low rank, with their oviducts opening at or near their second pair of antennae, and that their ova escaped at a period of exuviation, invested with an adhesive substance or tissue, which served to cement them together, probably, with the exuviae of the parent, to a supporting surface. In Cirripedes, we may suppose the cementing apparatus to have been retained; the parent herself, instead of the exuviae, being cemented down, whereas the ova have come to escape by a new and anomalous course" (1854, pp. 151, 152).

Darwin has thus clearly indicated that he was discussing evolutionary theory in regard to the supposed phylogeny of the cementing apparatus. The words "hypothesis," "unknown," "probably," "suppose" are all in keeping with such an interpretation. But on the other hand when he was comparing the representative specimens of the barnacles he studied, and classifying them as species in various categories as genera, sub-families, suborders or orders, he uses terms indicating the relative amounts and kinds of correspondence which he painstakingly observed and recorded in the monographs. Evolutionary suppositions are reduced to a minimum and in these descriptions Darwin makes frequent use of the terms appropriate for the classification of these barnacles. Without further lengthy quotations such terms are:

essential characters (*op. cit*, p. 247)

essential parts (p. 266)

essential point of structure (p. 374)

and many repetitions of these terms; and throughout Darwin uses the term "allied" to mean related in essential characters and the term "affinity" to refer to such relationships. Even the term "real affinity" (p. 565) is used to call attention to the essential rather than superficial or "analogical" characters and has no stated genealogical content.

To return to Ghiselin's commentary. "Darwin brought about a new way of thinking about morphological comparisons; he treated the archetype as a consequence of past evolutionary processes" (p. 109). But this new "thinking" does not appear in Darwin's "strictly systematic" work. To continue with Ghiselin. "The construction of archetype gave way to the reconstruction of common ancestors, and systematics became inseparable from phylogeny, a historical science of organic development" (p. 109).

My view is that Darwin's monographs do not warrant any such conclusion as that systematics became or should become "inseparable from phylogeny". But enough of the discussion centred about Darwin's monographs—though perhaps we cannot have enough of such consideration. At any rate let us now turn our attention to the principles of evolutionary systematics as revealed in the *Origin of Species*—which have been as much misunderstood as the monographs.

The theory. Evolutionary systematics as presented in the Origin of Species

Our discussion here will deal principally with Darwin's Chapter XIII in the first edition; Chapter XIV in the sixth edition. There is very little difference between them and what there is indicates that Darwin had taken a more practical and less theoretical view of his evolutionary theory of classification. For example in the first edition Darwin uses the term "community of descent" six times and the term "propinquity of descent" only once. In the sixth edition, "propinquity of descent" has been replaced by "community of descent".

Let us quote the sentences which contain these terms.

"I believe that something more is included; (in our classification than mere resemblance) and that propinquity of descent—the only known cause of the similarity of organic beings—is the bond, hidden as it is by various degrees of modification, which is partially revealed to us by our classification" (first edition, pp. 413, 414).

"I believe that this is the case, and that community of descent—the one known cause of close similarity in organic beings—is the bond, which though observed [sic] by various degrees of modification, is partially revealed to us by our classifications" (sixth edition, p. 205).

We have already called attention to the obvious typographical error noted also by Cain (1959a) in the case of the word "observed" which must surely have been intended to be obscured. But more important is the complete removal of the term "propinquity of descent" which apparently persisted through the first five editions, though used only once in each case. And important also is the qualification in regard to similarities—community of descent the one known cause of close similarity in organic beings. The earlier statements were clearly in error when they attributed all similarities in organic beings to descent. Darwin spent much time in explaining that many kinds of similarities did not reveal community of descent. In other words the changes made by Darwin bring his theory closer to his practise and provide a more accurate description of the real situation.

Now to the main principles of Darwin's evolutionary classification and their relation to his practise.

"All the foregoing rules and aids and difficulties in classification may be explained, if I do not greatly deceive myself, on the view that the Natural System is founded on descent with modification;—that the characters which naturalists consider as showing true affinities between any two or more species are those which have been inherited from a common parent, all true classification being genealogical;— that community of descent is the hidden bond which naturalists have been unconsciously seeking, and not some unknown plan of creation or the enunciation of general propositions, and the mere putting together and separating objects more or less alike.

"But I must explain my meaning more fully. I believe that the *arrangement* of the groups within each class, in due subordination and relation to each other, must be strictly genealogical in order to be natural; but that the *amount* of difference in the several branches or groups, though allied in the same degree in blood to their common progenitor, may differ greatly, being due to the different degrees of modification which they have undergone; and this is expressed by the forms being ranked undeı different genera, families, sections or orders" (p. 212).

Darwin then suggests that the reader restudy the diagram in his fourth chapter, excellent advice which ought to be followed by our present-day evolutionary taxonomists. The figure shows fifteen present-day genera which have descended from an earlier form. "Now all these modified descendants from a single species, are related in blood or descent in the same degree; they may metaphorically be called cousins to the same millionth degree; yet they differ widely and in different degrees from each other. . . . " So that the comparative value of the differences between these organic beings, which are all related to each other in the same degree in blood, has come to be widely different. Nevertheless their genealogical *arrangement* remains strictly true, not only at the present time, but at each successive period of descent (p. 213) . . .

"The representation of the groups, as here given in the diagram on a flat surface, is much too simple. The branches ought to have diverged in all directions. If the names of the groups had been simply written down in a linear series, the representation would have been still less natural; and it is notoriously not possible to represent in a series, on a flat surface, the affinities which we discover in nature amongst the beings of the same group. Thus, the natural system is genealogical in its arrangement, like a pedigree: but the amount of modification which the different groups have undergone has to be expressed by ranking them under different so-called genera, sub-families, families, sections, orders and classes" (p. 214).

Thus Darwin clearly states that the amounts of difference among the organisms studied have to be expressed in terms of the systematic categories chosen; and that this is perfectly consistent with the idea of their community of descent and even with a genealogical *arrangement* of these categories. But clearly there is no claim that any propinquity of descent is revealed by these categories nor any actual genealogy of species. "In classing varieties, I apprehend that if we had a real pedigree, a genealogical classification would be universally preferred; and it has been attempted in some cases" (p. 215) . . . "As we have no written pedigrees we are forced to trace community of descent by resemblances of any kind" (p. 217).

It is a great mystery to me why so many evolutionary taxonomists have misunderstood these statements and reached false conclusions about Darwin's practise of evolutionary classification, for his practise was strictly in accord with the theory inherent in the quotations we have selected. I have called attention to Simpson's error in his report "Anatomy and morphology; classification and evolution: 1859 and 1959" (1959) in which he states:

"What Darwin proposed was that classification should have a single theoretical basis; phylogeny or, in his own terms, propinquity of descent." (p. 300).

But Darwin proposed no such thing: nor did he practise classifying on any such basis. Ghiselin is equally in error when he states (1969, p. 89): "It has already been demonstrated that Darwin considered 'propinquity of descent' the basis of natural classification." It should hardly be necessary to quote further from Darwin's Monographs, but selected short quotations will prove that he was relying on the amounts of difference among the organisms to determine their categorical rank, just as he stated had to be done in the *Origin of Species.*

Under the description of *Balanus glandula* (1854, p. 266) he wrote:
Affinities—This species in general appearance closely approaches *B. crenatus* and *balanoides*, and it is related to them in many essential parts, such as in the opercular valves . . .

Pyrgoma monticulariae. (l. c. p. 373)
Affinities—"Although this species, as above stated, differs so remarkably in external appearance from the other species of the genus, and, indeed, of the whole family, yet the shell in no one essential point of structure materially differs from its congeners; and if we compare the opercular valves with those of the last three

species, we shall be struck with their close, yet graduated affinity. . . .

"If in *P. crenatiom* we were to remove the spur from the tergum (and it is much less developed in *P. dentatum*; and in *P. milleporae*, it is entirely absent) this valve would be almost identical with that of *P. monticulariae*. Under these circumstances I consider it impossible to separate the present species as a distinct genus."

Darwin's remarks in regard to the Family Lepadidae show again that his evaluation of the amounts and kinds of difference among its members were the basis for the selection of the appropriate taxonomic category. He was especially puzzled about Alcippe but though it was curiously modified he believed the remaining resemblances justified its inclusion in the family.

"But we shall presently find, when we come to Cryptophialus, that all the above difficulties, great as they are, are greatly enhanced, for Cryptophialus is certainly allied in a very direct and curious manner (in decided opposition to the remarks just made on special affinities) to Alcippe, and yet in all the more important parts of its organisation, and in its metamorphosis, it differs so fundamentally, that I have felt myself obliged to form not merely a Family, but a distinct Order for its reception" (l. c. p. 529).

Now to sum up. In his eight years' work on the barnacles, Darwin exhibited tremendous care in observation, comparison and evaluation of the complex similarities and differences among barnacle species. He drew many conclusions about the various grades of affinity which barnacle species show to each other, clearly explaining the nature of the similarities and differences upon which he based these conclusions. Evolutionary relationships were probably always in mind but he made no claims in regard to the genealogical relationships of his species which claims would have added no informational content to the classification. In most cases he avoided making any claims about the possible genetic or phylogenetic relationships of the various species. In other words his classification of the barnacles was almost, if not quite, "strictly systematic" in the sense that he used the characters available for comparison as he found them and grouped the species in accordance therewith. To me this is a real "Triumph of the Darwinian Method", and because his classification was based on facts it has stood the test of time. We hope that present-day taxonomists would more closely follow his example.

Systematics: Modern texts and theory

Our chapter on "Systematics, Evolution, and Phylogeny" is not to be considered as a mini-text on systematics, but rather an attempt to critically examine and appraise some of the basic assumptions and principles of present-day systematics, and to constructively criticize those which are unclear, uncertain or possibly unwise. There are several excellent texts dealing with the systematics of animals now available, the more commendable of which are:

Principles of Animal Taxonomy, Columbia Univ. Press, 1961 G. G. Simpson.

Taxonomy A Text and Reference Book, John Wiley & Sons 1967 Richard E. Blackwelder.

Principles of Systematic Zoology, McGraw-Hill, Inc. 1969 Ernst Mayr.

Classification and Biology, Aldine-Atherton Press, Inc. 1970 R. A. Crowson. (Originally Heinemann Educational Books Ltd.—London.)

We deliberately leave out of this list Hennig's *Phylogenetic Systematics* (1966), though we shall give it a measure of the criticism which it deserves.

To set the tone for our discussion let us quote a section from *The Nature of Systematics* (Blackwelder and Boyden, 1952, p. 33).

"Part III—Perspectives in Systematics

"It is time that universal taxonomy claimed its birthright. To collect, describe, name, compare, and group organisms by sound methods and principles is a great undertaking far from completion.

"As progress in taxonomy is made we shall then know better and more completely what kinds of organisms there are and where. From the systematized knowledge of animal kinds we may come to understand what evolution has accomplished; for it must be obvious to all that we cannot understand the nature of the process of evolution without knowing what kinds of organisms it has produced. It is the work of the taxonomist which provides the facts about the species and higher categories of animals from which are drawn inferences in regard to their phylogeny and the nature of the evolutionary mechanisms responsible. Without the work of the taxonomist neither the evolutionist nor the phylogenist nor anyone else would know what they were talking about or have anything significant to say.

"A new era in systematics lies before us. This era cannot be referred to as relating only to 'The New Systematics' of the past decade. It will, we hope, not be entirely either New Systematics or Old Systematics but SYSTEMATICS. In a manner analogous to the action of natural selection, the good characters of the Old and the New systematics will be selected and integrated into a functioning and harmonious organism. This SYSTEMATICS will attempt to classify the results of three thousand million years of evolution and to develop adequate principles for so doing. It must be frank about the limitations of available knowledge, and, while ever seeking to extend the limits of this knowledge, it will not wittingly confuse the facts of animal organization with theories and inferences in regard to their remote evolutionary history. In this spirit, a new generation of taxonomist may emerge, trained to be as rigidly scientific in methods as geneticists or biochemists, and to classify existing organisms as they are, or pre-existing organisms as the record indicates they were, but in neither case on the basis of inadequately documented inferences regarding their assumed ancestors. The accepted general guiding principles will be as universal as taxonomy itself, applicable to all kinds of animals, whether known only from the present, or whether, at the other extreme, belonging to a fossil lineage that can be determined with high probability. And above all, the grand object of classification will be realized, which is that organisms will be grouped in accordance with their natures so that the human mind may deal effectively with 'the myriad forms of life'."

Richard Blackwelder and I had at that time great hopes for a more practical and less theoretical taxonomy. After all, had not Simpson (1945, p. 17) said, "Classification is, above all, a practical problem". And in the same publication (p. 5),

"A phylogeny, even when perfectly known and universally accepted (which none is in detail), is not a classification, and the intricate relationship between phylogeny and classification leaves more than ample room for the exercise of skill and judgement and for differences of opinion."

"Phylogeny cannot be observed."

Now we must explain that the above brief quotations from Simpson may be misleading to those who are not familiar with the major parts of Simpson's "Principles of Classification . . ." (1945) and his "Principles of Animal Taxonomy" (1961). We consider Simpson to be the leader among present-day evolutionary taxonomists and there are vast areas in which we must accept his principles and statements as valid. Some criticisms we have already indicated—others will follow. They have to do mainly with the over-emphasis on theory to which phylogenetic taxonomy is so susceptible. There is apparently something about phylogenetic taxonomy which leads some of the members of its school to make the most absurd statements and accept the most fantastic beliefs!

We were not alone in our hopes for a more scientific and practical taxonomy. Thus Cain (1959b) said (p. 241):

"The development of precise procedures and the employment of non-human computers allows us to hope at last for the development of a taxonomic hierarchy with a really quantitative basis, a properly surveyed map of the diversity of living things instead of a highly subjective and distorted sketch map. The basis of these procedures is likely to be Adansonian. With the more general recognition of how much classification must—for lack of information—be entirely natural, there will be far less enthusiasm for the multitude of so-called phylogenetic trees that were produced so feverishly last century, when over-eager supporters of evolution thought that their case was incomplete unless every known form could be hung on some twig of the tree. It will be realized, for example, that the sort of arguments put forward by Hadži (1953) for his particular arrangement of the Platyhelminthes, Ctenophora and Coelenterata are on just the same logical footing as, and have no more validity than, the arguments used to uphold the more traditional arrangement. Hadži's work has great value at the present day, in reminding us how much (and how unjustifiably) under the spell of Haeckel are nearly all our text-books of systematic zoology."

And did we not have conferences such as that sponsored by the Systematics Association with the report of 1964 on *Phenetic and Phylogenetic Classification* where attention was knowingly given to matters of phenetics as well as phylogenetics? And was there not the report by Blackwelder which so clearly distinguished between taxonomy, the practise of classification and on the other hand, the study of speciation or of presumed phylogeny? To be sure the portrayal of a natural classification in Gilmour's sense as resulting from a thorough comparison of the characteristics of organisms and a consequent grouping into kinds was there but there was also such a view as that of Delevoryas who said (p. 20):

"In attempts to classify organisms, then, a system based on concepts of relationships and descent would seem to be the most worthwhile goal. If a system is not phylogenetic, it is really meaningless. If a system is not phylogenetic, it really cannot be judged 'better' or 'worse' than another that is not. It is really pointless to argue about the relative merits of various artificial schemes."

Such statements are completely incomprehensible to me. Human capacities to recognize basic correspondences among the characteristics of organisms (or of any other entities) and to group them in accordance therewith are among the most essential capacities we possess. There could be no effective thinking without them. We take a pragmatic view in this matter, much as does Meyer-Abich (1964). Crowson (1970, p. 1) has recently said, "Classifying things is perhaps the most fundamental and characteristic activity of the human mind, and underlies all forms of science" With this we fully agree.

It is so often believed as Delevoryas did, that phylogeny gives "meaning" to biological classification. Without becoming philosophical about this we should point out that any claim that common ancestry *explains* the characteristics of organisms has very limited validity. In the first place it does not explain the origin of any characters whatever. In the second place it does not even explain how, in terms of mechanisms, they are transmitted from generation to generation. Nor does it explain, in the third place, how some characters survive and others do not. The most that can be said is that knowledge of presumed common ancestry suggests where to look for partial explanations, that is to study the operation of the hereditary mechanisms concerned in character transmission in living organisms. But even if the world of life did not evolve, whatever kinds of organisms there were would have to be described, compared and classified, while a scientific study of whatever did determine their characteristics was being undertaken. After all, as we have said, systematics is universal.

Simpson (1961) and Mayr (1969) and others have reached decidedly mature views in regard to taxonomy and evolution. In fact Simpson now prefers to speak of "evolutionary" rather than phylogenetic taxonomy, for many reasons. In the first place, no classification can adequately express phylogeny, though it is held that it should be in accord with the author's inferences in regard to the past evolution of the taxa concerned, and in that more limited sense based on phylogeny. This practice leaves us with the difficult problems of inferring phylogeny where the fossil record is inadequate, which is the commoner case taking the animal kingdom as a whole. It also requires that distinctions be made among all the complications of divergence, parallelism, and convergence in the history of taxa. The fact is that "evolutionary taxonomy" is not achieved unless this can be done; and the term is not really correct for it is nearly a universal practice among evolutionary or phylogenetic taxonomists to ignore the results of convergent evolution where it has apparently occurred. Evolutionary taxonomy in practice, therefore, means divergent evolutionary taxonomy only. No matter how much the nature of the organisms has changed as a result of convergent evolution, the taxonomic values of such convergent characters are markedly and deliberately reduced in the resulting classification. To a certain extent, therefore, "the quality of the goods as delivered" (Bather, 1927) is ignored in such classifications, and their practical usefulness or value for prediction is also reduced.

We have spoken about the grand object of classification being everywhere the same in all branches of human knowledge. That is, the purpose is to group in accordance with nature, objects whether animate or inanimate; also ideas, or any other parts of human knowledge. But organisms do have history whether it can be known or not and inanimate objects may have a very different kind of history. There is no inherent code of instruction and no reproduction in the biological sense among them. Mayr's (1965) recent discussion of "Numerical Phenetics and Taxonomic Theory" bears directly on this matter. He states (p. 77) that the evolutionary taxonomist "rejects categorically the suggestion that there is no difference between classifying organisms and inanimate obejcts". Mayr's whole discussion is most clear and helpful in regard to this particular aspect of classifying, and to many others as well, and a careful study of his discussion is strongly recommended. But let us examine a little more carefully the comparison and classification of inanimate objects in contrast with the methods of comparing and classifying organisms.

Classification of inanimate objects compared with the classification of organisms

The statement that organisms should be classified using the same basic principles as are used for inanimate objects has been disclaimed by those who support a phylogenetic or evolutionary classification, most recently by Mayr (1969). The assumption seems to be that it is proper to classify inanimate objects in accordance with their nature or uses but that organisms cannot be and should not be classified so "objectively". Simpson (1961) has often spoken of the "intuitive" nature of biological classification, maintaining that even with Richard Owen's criteria of homology, the operation of judgment is essential. On the other hand, he claims, wrongly I believe, that the recognition of common ancestry, or determination of "propinquity of descent", is less subjective than the appraisal of the essential structural similarities which Owen stated to be the criteria of homology. I have (1960) attempted to show the fallacy in this line of reasoning, indicated by the necessity of establishing the existence of structural or other correspondences, which is the first step in the analysis which could lead to an inference of common ancestry. There is no other route to such a conclusion and if the recognition of essential structural correspondences is "subjective" as Simpson claims, the inference of common ancestry based thereupon cannot suddenly become objective on the grounds that there may have been common ancestry.

When we examine the operation of classifying more carefully, whether for organisms or inanimate things, we see that even for the latter, judgment is required. Among the various attributes possessed by such objects as tacks, nails, screws, and bolts, two major kinds of characters are shown in their present natures; composition and form. Which of these is the more important, i.e., useful for a "general purpose" classification? Tacks may contain copper or steel; they are still tacks. So with nails—soft iron, galvanised, coppered, but all are nails. Obviously, composition is believed to be less important than form in such a general purpose classification. So it is with the other articles mentioned; the possession of a thread distinguishes all manner of screws and nuts and bolts from nails and tacks. Further, with regard to form, the thread is more important than the shape of the head which may be round or flat. But the necessity of a functional two-piece assembly separates the nuts and bolts from lag-screws. The point of all this is that there is a weighting of characters required in any useful classification of inanimate objects, which in turn requires judgment rather than intuition. Any numerical taxonomy could come up with a relatively objective evaluation of these similarities, but this would not necessarily result in a more useful classification.

Wherein is there then a valid distinction between the methods of biological classification and those suitable for inanimate objects? Not in the requirement of judgment, but rather chiefly in regard to the question of ancestry, of history and here we are confronted with the controversy between phenetic and phylogenetic or evolutionary classifications. This controversy has continued for a century and yet we believe progress is being made toward greater understanding of the true nature and goals of taxonomy and of systematics, both new and old.

We may refer to the operation of grouping organisms on a chiefly phenetic basis as a "present nature" method to distinguish it from avowedly phylogenetic methods. Though the phylogenists must work through phenotypes to get to their inferences in regard to genotypes and ancestry, they appear to be attempting to discount these phenotypes or to "correct" nature, whenever they believe that the ancestry was different than the present natures of the organisms would indicate. This practice has led to some confusion and many changes in classification as knowledge of nature and probable ancestors has increased.

Mayr (1965) has recently clarified the relations between evolution and phylogeny in a very helpful way, but first let us explain the "present nature" basis for classification of organisms so it need not be misunderstood.

The present nature method deals with organisms as they are, or in the case of fossils as they were, on an admittedly but partially phenetic basis. Nevertheless, it is not the same method as would serve for the classification of inanimate objects in several respects relating to their history. The present nature method:

1. Does not ignore sexual dimorphism, nor any kind of polymorphism

which exists within species, nor does it ignore individual ontogenies.

2. Distinguishes wherever possible between genetic and non-genetic characters and gives the greater weight to the former in accordance with their greater consistency.

3. Does not ignore convergent characters, yet does not give them more than their due in the final classification. In other words, it seeks to avoid failure to recognize "the qualities of the goods as delivered", which in our pragmatic view of the uses of classification is as great a crime as to classify on phylogenetic inference in spite of the present nature of many characters.

Mayr's recent comments (1965, 1969) have a great relevance in this matter of the interrelationships of phylogeny and classification. He points out that "Phylogeny is characterized by two basic evolutionary processes, the branching of lineages (cladogenesis) and the subsequent divergent evolution of daughter lines". Schools of taxonomists differ in their relative emphasis upon these aspects of evolution. Some take the one-sided approach of using the splitting as the only process in evolution of significance for taxonomy. According to Mayr, this is true of Copeland and Dillon, and typical representatives of this school, claimed to represent the true phylogenetic approach, are Hennig (1950, 1966) and Kiriakoff (1959). Mayr (1965) states plainly and with right that it is not the true phylogenetic approach since it ignores the amounts of evolutionary divergence which followed the splitting. In other words, this procedure clearly ignores the qualities of the goods as delivered and must therefore fail to produce a general purpose classification. "It is misleading if the cladists call themselves the phylogenetic school, since they ignore a large part of the phylogenetic process." On the other hand, pheneticists could be equally in error if they ignore the differences between genetic and non-genetic characters, or matters of dimorphism and polymorphism within species. This they do not do. As far as the numerical taxonomists are concerned (Sokal and Sneath, 1963), there is an operational limitation in their exclusion of the relatively conservative characters possessed by the organisms they classify, viz., the exclusion of all characters which do not vary among the individuals and taxa studied. Again, this procedure ignores certain of the qualities of the goods as delivered and to that extent fails to meet the ideal requirements of a general purpose classification.

The present nature method, then, attempts to achieve a general purpose classification, using all kinds of biological characters and weighting them on their consistency or conservatism or on the basis of their systematic distribution with obvious correlations of attributes into patterns; but never making claims that such a classification reveals phylogeny adequately, much less that it is based "on" phylogeny. It could never confuse genealogic with genetic relationship, nor claim that the study of existing organisms alone can provide a truly phylogenetic classification. To this extent, it is a more scientific approach to taxonomy than most others and is worthy of serious study and development.

Mayr (1965) speaks of the important distinction between genetic and genealogic relationships. It is unfortunate that, following Darwin, so many have failed to make this distinction. Darwin tried to make clear that the natural system should be *"genealogical in its arrangement"* which is not the same as a genealogy. A genealogy is, properly speaking, a pedigree record complete as to individuals and generations and not even the most complete fossil lineage meets these specifications. To equate phylogeny with genealogy is really in most cases impossible and a more critical view of these concepts is long overdue. According to Meyer-Abich (1964, p. 60), "There is no phylogeny of species, but a phylogeny of the typological characters of the species," and "Phylogeny is not the genealogy of the species, but exclusively the history of all biological characters which exist today, in the past, and may possibly exist in the future" (p. 66).

To clarify these points a little further, Mayr cites the case of the evolutionary history of the birds and crocodiles. The cladists' philosophy forces them to put birds and crocodiles in one taxon, the rest of the reptiles in other taxa. But obviously the crocodiles are still closer to many of the other reptiles, in characters of genetic determination, and therefore in genetic relationship, than they are to birds. Here again a phylogenetic classification of the cladistic type, misrepresents the natures of the final products and fails to achieve the usefulness of a present nature or general purpose classification.

We have spoken about genetic versus genealogic relationships which may be very different, depending upon the rates of evolutionary change in the genotypes of the diverging organisms. Such genotypes have been inferred from the phenetic characters of all pre-existing organisms. As to existing organisms, detailed knowledge of genotypes has been obtainable only within the limits of interfertility among the individuals concerned. Obviously this knowledge is very restricted and is incapable of giving evidence regarding remote genetic relationships. It is true, as I have explained (Boyden, 1934, 1943, 1947) that the mechanism of homology is the mechanism of heredity, and therefore we may assume genetic relationship as far as the limits of homology extend. Many students of morphological homology believe it best to use the criteria for homology which Owen specified (1843, 1847, 1848) and to limit it inclusiveness to organisms of the same type. Obviously this gives no answer to the evolution or phylogeny of the types themselves, or to the problems of genetic relationship between them. Recently the limitations of classical genetic methods have been extended by studies of homology between the DNA and RNA of diverse organisms. This new development in "Molecular Taxonomy" seemed promising for the study of some of the more remote genetic relationships and should be carefully considered. Let us discuss it briefly before we pass on to consider the place of genetics in modern systematics.

Taxonomic implications of DNA–RNA pairing

There has recently appeared a series of reports, describing simple, effective procedures for measuring the relative homology or complementarity between the DNA and/or RNA of diverse organisms. Instead of the difficult density gradient procedure for detecting enlarged (hybrid) DNA–RNA compounds, the single strands of DNA resulting from heat treatment are trapped in suitable concentrations of agar gel and subsequently tested for combination with complementary sheared strands. The methods described by Bolton and McCarthy (1962) have been shown to be very useful. The results obtained first in bacteria and related phages and then among bacterial species, led to this conclusion; "clearly, the method here employed appears to have potential for exploring quantitatively the genetic relatedness among species."

Further reports develop and bring to realization some of this potential, and begin to consider the broader biological implications of the results. Thus the report of McCarthy and Bolton (1963) begins with this paragraph:

"The presence of genes in common may be taken as a guide, not only to taxonomic relationships among organisms, but also to probable evolutionary relationships. However, reproductive isolation of distantly related forms precludes the use of the usual methods of genetics for the determination of gene similarities".

The report then continues to describe the relative extents of hybridization between a number of DNA agar preparations and the RNA and DNA fragments of different bacterial species, and emphasizes the usefulness of such studies, especially among bacteria, for understanding evolutionary relationships "where there exists only the faintest paleontological record and the simplest of all ontogenetic processes."

More recently we have the report by Hoyer, McCarthy, and Bolton (1964) which extends the methods to higher organisms, and reviews the procedures employed. To those who have long realized the narrow limitations of classical genetic methods for the study of genetic relatedness, their results are a great revelation. We quote from their conclusions:

"It is clear from the results presented that there exist homologies among polynucleotide sequences in the DNAs of such diverse forms as fish and man. These sequences represent genes which have been conserved with relatively little change throughout the long history of vertebrate evolution."

The details should be studied in the original reports, but we must mention one further observation, viz., "that the genetic diversity among families of bacteria is relatively greater than the genetic diversity among all the major vertebrate classes".

The above is a very brief summary of one kind of approach to the molecular taxonomy of living organisms. Though its present employment is relatively limited, we can begin to see some of its merits and some of its limitations.

There is no question in our mind in regard to the conclusion that it can provide evidence of "genetic relatedness" among existing organisms. Like systematic serology, however, it is practically limited to such organisms. Though it extends the study of genetic relatedness far beyond the limits of hybridization applicable in classical genetics, it has no time axis and therefore cannot supply any verifiable phylogenies. All inferences about the tree, or trees, of life are based on the genetic codes of their terminal twigs. Important as this approach is, it is quite unrealistic to expect it to provide answers to a host of questions in regard to the remote ancestry of the present world of life.

The next question is concerned with the possible relationship between the extent of homology revealed by DNA–RNA pairing and the taxonomy of the organisms developed under the direction of such genetic codes. Is genetic relatedness so revealed directly convertible into a useful classification? We question the validity of the idea that ". . . the total genetic potential of an organism is represented in DNA . . ." (Hoyer, McCarthy, and Bolton, 1964, p. 959). Even for all those organisms which have a DNA code, the total genetic potential of DNA by itself is zero. It takes at least the major parts of a cell to produce another cell, and no one has ever succeeded in getting an organism even from an entire sperm!! Only living organisms have genetic potential. Besides, the genetic code may include some parts of the organization of the cytosome, parts such as certain regions of the cortex of *Paramecia*, which are independent of DNA instruction (Sonneborn, 1963, 1967).

We do not believe at the present time that there can be any simple and direct conversion of indices of genetic relationships based on DNA homologies into general purpose classifications. Our reservations are based chiefly on the following facts and considerations:

1. Gene mutations, even those resulting in drastic alterations of the phenotype, may not alter the process of pairing among the alleles so produced; yet such genes and the chromosomes which include them would be considered "homologous" by the DNA–RNA hybridization procedure.

2. There is no simple proportionality between the amount of similarity or "homology" in the genetic codes and the amounts of similarity in the phenotypes of the developed organisms.

a. So far as we know, polymorphic organisms with metagenetic life-cycles such as Hydrozoa, Scyphozoa, Trematoda, and Cestoda, possess but single codes in each species. However, the phenotypes expressed in the life-cycles may be quite diverse. It could be expected that the DNA–RNA hybridization index would give one relative value for these species, whereas the phenotypes of the hydroid stages might lead to a different taxonomy than the pheno-types of the medusoid stages, etc. But how would one then classify species having metagenesis, such as species of *Obelia*, with species lacking metagenesis, such as *Hydra*, or any one of a host of sea anemones? Obviously, more than the genetic codes must be compared if any useful classification related to the natures of the organisms is to be obtained.

b. Long ago, De Vries stated (1910), "Systematic relationship is based on the possession of like pangens. The number of identical pangens in two species is the true measure of their relationship." Some fifty years later, the students of DNA–RNA hybridization came forward with a basically similar proposal. For pangen, the word gene is substituted and for "like" genes, genetic homology is used. But the principle involved suffers from the same limitations now as then. In the first place, it is not the amount of genetic homology, or the number of like genes, but rather the proportion-ate amount of homology to the total code which should be used if a correlation between the similarities of the codes and the organisms is desired. In the second place, it is assumed in each case that a given proportion of code similarity will give the same amount of phenetic similarity. This could only result if the genes evolved always at equal rates and had equal effects upon the natures of the organisms concerned. Neither of these assumptions appears to be true.

3. The more recent discovery of relatively large amounts of duplica-tion of genes in the DNA strands (Britten and Kohne, 1968; Britten, 1969) has raised some further uncertainty in regard to the taxonomic implications of DNA pairing. According to Britten (p. 209) the measurements made of the DNA sequences in common among species include the bulk of the repeated sequences. The ontogenetic function of such repeated sequences is still unclear but that they play an important part in the measurement of the relative amounts of DNA–RNA homology among species is clear, and makes any direct translation of such knowledge into bases for determining taxonomic arrangements most unpromising.

For the time being, we can only say, therefore, that we find the approach to molecular taxonomy of Bolton and colleagues most interesting and stimulating, and we look forward to its further extension. Especially, we would like to see it applied to studies of the evolution of the lower Metazoa, where previous methods of study have led to considerable diversity of opinion (Dougherty, 1963); also to the study of organisms which may have undergone paedomorphosis in evolution (De Beer, 1958) where presumably small genetic change has produced large phenetic alterations. Genetic relatedness as indicated by DNA–RNA pairing could help to solve some of the problems of ancestry in these organisms, but whether this would justify any drastic change in their classifications is not at all certain. And the limitations of this knowledge must be constantly borne in mind.

As we have said, genetic relationship is of many kinds and degrees and therefore has no single kind of precise meaning, other than the general requirement of common ancestry. The phylogenists, of course, make much of assumed common ancestry and tend to ignore or belittle the amounts and kinds of similarity in the somatic character expressions. For the cladists there may sometimes be a claim that morphologic similarities are of secondary importance. According to Hennig (1966, p. 74), "the measure of phylogenetic relationship is the relative recency of common ancestry." And Hennig also quotes Zimmerman (1931) with approval, " 'The relative age relationships of the ancestors is the only direct measure of phylogenetic relationship.' "

The criticism is obvious. If "relative age relationships" are the only "direct" measure of phylogenetic relationships, then in truth there are no such direct measurements at all!

Hennig (1966) admits as much when he states (p. 94) "Naturally, in determining homologies we are limited to erecting hypotheses—such as that particular characters a, a', a" belong to a phylogenetic transformation series." And he discusses at some length (p. 88ff) "The Rules for Evaluating Morphological Characters as Indicators of Degree of Phylogenetic Relationship." In this discussion there is an eruptive evolution of words which hardly seem improvements over existing words. Thus primitive character states are called *plesiomorphous* and the presence of such characters in different species is *symplesiomorphy*. Derived character states are called *apomorphous* and the presence of such derived characters in different species is called *synapomorphy*. And on p. 92, he states that "When we distinguish different conditions of a character (a, a', a") as plesiomorphous and relatively apomorphous, we do not take the magnitude of the differences between these conditions into account in any way."

The very complicated discussion of the primary and "accessory criteria" dealt with is illustrated with many beautiful line diagrams supposedly illustrating the evolutionary processes involved. Can anyone be so naive as to believe that these adequately represent the evolutionary processes involved? Or that they have any direct bearing on the major and practical concerns of taxonomy?

To the extent that the cladists depend on the relative age relationships of the species studied they are ignoring a major part of the evolutionary process, and to the extent that they use morphological characters to aid in determining symplesiomorphy vs. synapomorphy, they are circumventing the major part of the taxonomic process which is to group organisms in accordance with their observable characters with minimum assumptions in regard to them. Can it be that cladists have some inward reservations in regard to the validity of their phylogenetic determinations and therefore turn to the geologists and physicists for relative ages of branching as simple solutions to the problems?

Reservations or not, the determination of the time of appearance of organisms as fossils is subject to some errors and is hardly accurate enough to serve all taxonomic purposes. There is one particular source of error which I have not seen mentioned before. How long must a new species be in existence before any of its members become fossils? How many fossils must there be before any is likely to be discovered. Obviously the time intervals could vary enormously depending on the nature of these new organisms; the existence of skeletal structures, the habitat, etc. The point is that the answers to these questions are not only unknown—but even unknowable!

We raise one further matter in connection with phylogenetic taxonomy in its relation to genetic correspondences. Within the limits of interbreeding, genetic relationship of some kind is directly determinable. Beyond such limits (whether they are precisely the same as taxonomic species limits or not) there is no direct measure of genetic relationship and any claims that particular classifications reveal such relationships can be approximations only. The fact is that genetic relationship is of such a great range of kinds and degrees that any translation of such relationship into taxonomic categories can only be of very limited usefulness in reaching a natural classification—i.e., one in accord with the present nature of the organisms.

Now when Mayr (1969, p. 77) states:

"When we classify organisms, classification by phenotype is only the first step. As the second step we attempt to infer the genotype, the evolved genetic program, which has a far greater explanatory and predictive value than the phenotype. The phenotype is susceptible to all sorts of irrelevant similarities, and it is only the analysis of the inferred genotype which permits us to determine what similarities in the phenotype are due to convergence and what others are an expression of the ancestral genotype."*

Here is the crux of the matter of the difference between the phylogenetic and present nature bases for classification. The former does not classify organisms as they have evolved, but inferences as to what they should be! The phenotypes are circumvented and inferences in regard to genotypes are raised to decisive levels in grouping. We believe the whole theory of denigrating the somatic expressions is unsound, for the following reasons:

1. Organisms live and die by their somas. The whole office of the germ plasm is to produce and reproduce viable somatic characters. Natural selection can act directly only on such expressed characters, or can eliminate germ plasms which fail to produce and reproduce them.

As to the alleged "all sorts of irrelevant similarities" we question that natural selection would really pass on the capacity to produce such similarities. The report by A. J. Cain "The Perfection of Animals" (*Viewpoints in Biology* 1963, **3**, 36) could lead to a better appraisal of the importance of the phenotypes in evolution and classification.

As to inferences in regard to the genotype being able to distinguish what characters are convergent and what are not, it should be pointed out that persistent convergent characters are as much due to heredity as divergent characters! The important question for the phylogenetist is not whether the characters are inherited or not, but whether the

* From Mayr, E. *Principles of Systematic Zoology.* Copyright 1969, McGraw-Hill Book Co. With permission.

somatic expressions have been present in a continuous succession of ontogenies since their origin. Evidences to support such a conclusion are often minimal or lacking and why such a conclusion should be decisive in regard to classifying the final products of evolution is apparent only to the avowed phylogenists.

Then there is the further matter of the treatment of correlated characters. These are characteristic of the present nature of many organisms. Mayr (1969, p. 221 ff.) states that a strict distinction must be made between functionally correlated characters and phyletically correlated characters. The former are to be given low weight because they are "redundant"! In other words characters, known to be functional and inherited, are to be given low weight in classification because there are too many of them!! Here is a striking illustration of the phylogenist's practise of belittling the natures resulting from the evolutionary processes and raising to the level of decisiveness, inferences in regard to the amounts of similarity in the genotypes. But we have already pointed out that relative amounts of similarity in genotypes— even such as the proportions of "like genes"—have no simple relation to the present natures of the organisms compared and to be classified. Correlated characters, functional or not, characterize organisms and organisms should be classified accordingly. Natural selection has passed on such correlations and they are an inherent part of the nature of the organisms possessing them.

What then is this matter of "predictivity" being so much greater on the basis of alleged phylogenetic classifications? Would ignoring "redundant" characters enable anyone to predict how an organism lives? On the contrary, it could be very misleading. The best basis for predictivity in regard to the way an animal lives, what it can and cannot do, would be a general purpose classification based on all characters, even variable ones. If by predictivity is meant guessing as to how the species will evolve in the next half-million years, the study of past evolution would give a better basis, but what has this to do with classifying organisms now? The chief virtue of the present nature method is that it attempts to classify organisms not inferences. It is as much an evolutionary classification method as any other, witness Darwin's *Monographs on the Sub-class Cirripedia*. And it does not obscure the quality of "the goods as delivered" as the phylogenists are determined to do. We repeat, the first essential in systematic zoology should be "systematic thinking". (Boyden, 1947, p. 668).

Blackwelder (1964, 1967) has consistently stated that the study of speciation is not an essential part of the procedures of taxonomy, which should be concerned and for the most part is concerned with the study of kinds of organisms.

Let us continue by quoting from a report by Blackwelder and Boyden (1952) concerning "The Nature of Systematics" (pp. 29–32).

"Part II—The Aims and Tasks of Systematic Zoology

"Systematics draws upon all sources of data about organisms for information in regard to their natures, and it must therefore concern itself directly or indirectly with the validity of all methods, principles, and practices which are used in characterizing these organisms and arranging them in accordance with their resemblances and differences. Those branches of zoology which contribute to systematics the facts of comparative anatomy, comparative physiology, and ecology have rather obvious and well understood aims and tasks. Their goals are to make known all the facts about the structures and functions and daily living of animals and to develop effective and suitable procedures for the study and reporting of such facts. But those branches of zoology and systematics which are primarily concerned with the systematization of this knowledge, i.e., with the characterization and grouping of kinds of organisms and the choice of biologically significant characters for doing this, require further treatment.

"It is a strange fact that neither the nature of systematics nor its real aims and tasks are correctly understood in these times. For this situation the responsibility must be placed primarily upon the systematists themselves who have, for nearly a century, consistently misorientated and belittled the scope of their own work. Whereas, prior to the publication of Darwin's *Origin of Species*, the work of systematists was viewed as collecting, describing, naming, comparing, and grouping organisms, it has since been more commonly viewed as having one and only one major aim, which is implied in the statement that 'the goal of taxonomy is the expression of phylogenetic relationships.' At first this too-close identification of the tasks of the systematist with those of the student of evolution was revealed in the creation of numerous all-embracing phylogenies from Monera to man, or as the modern phrase puts it, 'from Amoeba to man', and phylogenetic speculation was rampant. It was an inevitable result of many decades of ancestor-hunting of this kind that inferences about ancestry came to be considered more important than the facts of comparative anatomy, as is clearly witnessed in the still prevalent definitions of homology which state that homologies are resemblances 'due to common ancestry' and omit all other specifications as to the kinds and amounts of such correspondences. A moment's serious thought should make clear that inferences regarding ancestry must be derived in the first instance from the amounts and kinds of structural or biochemical correspondence, the latter referring to structure on the molecular level of the proteins characteristic of the body, and that to define homology as 'resemblance based on common ancestry' is to misunderstand the nature and limitations of the available knowledge.

"This misorientation of systematics is revealed also in recent writings characteristic of 'The New Systematics'. Thus Huxley (1940) has stated that "Fundamentally, the problem of systematics regarded as a branch of general biology is that of detecting evolution at work". Here the scope of systematics has apparently been reduced to the narrowest possible dimensions—that of "detecting", which presumably means witnessing the actual process of evolution in the present, or perhaps analyzing its immediate results. Nothing is said in this statement about the grand tasks of collecting, describing, naming, comparing, and grouping the results of one thousand million years of evolution. Actually, *The New Systematics* is broader than Huxley's quotation would imply, and it seems evident from the rest of his writings that he intended no such limitation on the scope of systematics. But, on the other hand, much of the recent writing in the field of the new systematics does seem to accord with the specification of "detecting evolution at work". For instance we may turn to the recent definitions of species to be found in Dobzhansky (1937) and Mayr (1942). In Dobzhansky's *Genetics and the Origin of Species* we find the proposal to define species "as that stage of evolutionary process at which the once actually and potentially interbreeding array of forms becomes segregated in two or more separate arrays

J

which are physiologically incapable of interbreeding." And we find also a similar concept in Mayr's definition, "Species are groups of actually or potentially interbreeding natural populations, which are reproductively isolated from other such groups". In these writings it is evident that a close fusion of the detail of evolution in the present with the business of systematics has been attempted and that the great task of "arranging the myriad forms of life" (Simpson, 1945) has been inadequately represented.

We quote from Simpson in regard to the nature of taxonomy.

"The science of arranging the myriad forms of life is taxonomy. . . . A formal classification of animals is a part of taxonomy, but only a part when 'taxonomy' is used in its full sense, and it is also a practical application of some particular set of taxonomic principles, of which many different sets are possible and useful." [Taxonomy in the "full sense" as defined by Simpson is equivalent to our systematics. Confusion due to the use of one term in both broad and narrow senses can easily be avoided by our usage.]

" 'Taxonomy' (that is, systematics), is at the same time the most elementary and the most inclusive part of zoology, most elementary because animals cannot be discussed or treated in a scientific way until some taxonomy has been achieved, and most inclusive because taxonomy in its various guises and branches eventually gathers together, utilizes, summarizes and implements everything that is known about animals, whether morphological, physiological, psychological, or ecological.

"Classification consists of grouping things according to their characteristics or properties, placing them in a system of categories, and applying a designation to each group thus established. It is obvious that Simpson's statements are greatly different from Huxley's characterization of the business of systematics as "detecting evolution at work", though these two authorities probably agree much more in regard to the nature of systematics than the quotations given indicate. However that may be, the relation between certain of the subdivisions of systematics, such as taxonomy, phylogeny, and classification must be clarified. In the first place the relations between taxonomy and classification are intimate and inseparable, so much so that it is convenient to refer to a taxonomist as one engaged in the description, naming, and arranging of organisms. But a taxonomist may or may not be also a phylogenist, for the nature of the relation between taxonomy and phylogeny is of a less immediate kind. Let us speak of the taxonomist as the person engaged in describing and grouping organisms and of the phylogenist as the person who attempts to trace their ancestry. If taxonomy is to be considered scientific in its methods, the bulk of the knowledge concerned with it must be observable and verifiable. But not only must the facts of animal organization used by the taxonomist be "hard facts", but the grouping of organisms based on such facts must be a logical process.

Now the grand object of classification everywhere is the same. It is to group the objects of study in accordance with their essential natures. Any scheme of grouping based on other attributes which results in placing together unlike objects and separating like objects will defeat the major purpose of grouping. But in biology, since Darwin's *Origin of Species* appeared, biologists have substituted for this grand object that of "expressing the phylogenetic relationships" of organisms, a substitution which has introduced endless confusion into taxonomic theory and practice. The complex relations between taxonomy and phylogeny have been discussed by many authors, including Huxley (1940) and Simpson (1945). Let us attempt to clarify the relations between them by listing the aims and tasks of those who work in each field.

The aims and tasks of the taxonomist.

1. To participate directly or indirectly in the great task of collecting representative samples of existing and pre-existing organisms.

2. To describe these organisms in all their essential characters as accurately and as fully as possible.

3. To compare these organisms objectively and quantitatively.

4. To develop a set of principles in regard to the choice and relative importance of those characters which will permit a grouping of the organisms possessing them in accord with their essential natures.

5. To group organisms on the basis of the characters chosen into a series of more and more inclusive categories, and to give appropriate names to the categories needed.

6. To develop keys and other devices for the recognition and

identification of organisms and to help non-systematists learn the names and relationships of the organisms with which they deal.

7. To help clarify the place of systematics in general biology and to give some thought to the improvement of the training of general biologists and professional taxonomists.

For comparison with these aims and tasks of the taxonomist we list those of the phylogenist, who seeks to trace the broad outlines of racial descent. This is also a complex and difficult task, a necessary motif for the taxonomist according to Parker but not so considered by us. The challenge of bringing order out of chaos, of properly arranging "the myriad forms of life" in accordance with their natures, is quite a sufficient motivating force.

The aims and tasks of the phylogenist.

1. To study the operation of genetic mechanisms in order to describe and understand the nature of the evolutionary processes in the present.

2. To analyze the genetic mechanisms responsible for homologous correspondences and thus to determine the ways in which somatic character expressions may indicate common genetic mechanisms.

3. To distinguish among traits of all kinds those which are conditioned mainly by heredity.

4. To determine which characters are primitive and which are specialized and to determine also whether the characters have diverged, run parallel, or converged.

5. To distinguish clearly fact from theory in phylogeny and to attempt to determine whether present knowledge in regard to heredity and variation can provide an evolutionary mechanism which could, if sufficiently extended, account for the production of all the types of animals which now or formerly inhabited the earth.

This listing of the aims and tasks of the taxonomist and the phylogenist separately should help to clarify the relation between them, a relationship which has also been effectively discussed by Huxley (1940) and Simpson (1945). "Phylogeny cannot be observed." This quotation from Simpson is characteristically frank and indicates that conclusions relating to phylogeny are based on inferences from observed facts regarding the observable characters of organisms and their space–time distribution. On the other hand the characters of existing organisms can be known and the characters of pre-existing organisms can be known, in part, depending upon the extent of the preservation of the ancestral characteristics. Phylogeny can at best be but fragmentary as to the characters of any organisms and as to the proportion of organisms having a sufficiently adequate fossil representation to make a phylogeny possible. On the other hand taxonomy must be universal. Conservative estimates indicate that there have been at least a thousand million years of evolution on the earth,* and of this period less than half is documented with fossil remains, and in a very spotty and fragmentary manner at that. Furthermore, this documentation is not only a fragmentary one for most animal groups, but the later stages rather than the early formative stages are the ones which are preserved and which represent most commonly the hard parts only. Many groups of soft-bodied animals are practically unrepresented in the record and their origin cannot be known, though the present nature of their existing representatives can become known in whole and that of their pre-existing representatives may be presumed.

To continue to subscribe to the view that "the goal of taxonomy is the expression of phylogenetic relationships" or even that a classification must be based on phylogeny is not now and never has been justified. We grant that the classification of the results of evolution may have "a phylogenetic background", and that in general the present nature and phylogenetic methods will, if correctly applied, yield concordant results. But the systematic method based on the present natures of the organisms grouped must be given the precedence over the phylogenetic method (1) because of the universality of the former and (2) because of the greater objectivity and verifiability of the facts used in the present nature method. Furthermore, the only reason why a phylogenetic method may be useful in systematics is that, where the evidences permit, it will give results in accord with the present or essential nature method. If the phylogenetic method, which groups according to presumed ancestry, should give results greatly unlike the present nature method, the former would be a useless method for the real and necessary purposes

---

* Recent fossil discoveries indicate that this estimate should be increased to three thousand million years.

of systematics. However important the phylogenetic method may be for the student of evolution, therefore, it must be given second place by the taxonomist, for his great aim must be to group existing organisms on the basis of their natures and by means of the characters which show what the organisms are now, and to group pre-existing organisms on the basis of what they were as indicated by their fossils, and not to group any organisms on the basis of what their ancestors were or may have been.

Now we should like to conclude this section of our discussion with the help of a recent report by Darlington (1971). It is pointed out in this excellent report that the current taxonomic literature is of two different kinds, viz., one kind is exemplified by many papers in *Systematic Zoology*, and these are mainly theoretical or experimental and often mathematical. The other kind is represented by the great volume of work reported annually in the *Zoological Record* and this kind is practical, concerned with classifying actual organisms by more or less conventional methods. Darlington speaks of the gap between the two kinds of biologists who work with these different approaches to taxonomy and endeavours to promote an exchange between them, by his careful analysis of their works. This is most encouraging for the future.

As to the importance of the practical worker and his reports we must point out that Blackwelder has for a long time emphasized practical taxonomy and pointed out its significance for all the workers in biology or elsewhere. His reports of 1962, 1964 and the text of 1967 all pay tribute to the practical worker and his essential contributions to the whole of biology including evolutionary studies. And Waldo Schmitt (1954) has written a valuable report entitled: "Applied systematics: the usefulness of scientific names of animals and plants," a report which makes clear how such names lead to essential information needed in the solution of many of the problems met with in agriculture, and medicine, and all phases of practical biology.

To return to Darlington. His entire report is worthy of careful study. A few quotations may be suggestive and we hope will lead some of the theorizers to examine it.

"It is a misfortune that theoretical taxonomists have often failed to appreciate the new elements in 'conventional' taxonomy and have also failed to look for workable compromises between current practice and new theory. This failure has, I think, retarded the evolution of taxonomy as an efficient, useful, respected subscience" (p. 342).

"As a biologist familiar with birds and some insects in their habitats, I am convinced that *biological species* do exist. . . . But biological species are complex populations or groups of populations. They cannot be measured as wholes or precisely; they can only be sampled. And the samples do not show all the limits or all the variability of the whole populations" (p. 344).

"Some practical taxonomists are moving toward reality, by using appropriate modern resources and procedures to define species and to make classifications more consistent with situations in nature. On the other hand, some theoretical taxonomists are moving away from reality, away from living organisms and natural situations toward purely numerical and mathematical concepts. One step in this direction is to replace biological species with OTUs. Another is to work with second-hand or hypothetical cases. And another is to manipulate the OTUs without reference to their biological origin or biological significance" (p. 346).

Darlington continues with sound advice in regard to the "Rules of the trade" (p. 347) followed by a discussion of Specifics: "an example of modern practical taxonomy" (p. 348). The example discussed is C. H. Lindroth. The ground-beetles of Canada and Alaska (Opuscula Ent. (Lund) 1961, 1963, 1966, 1968, 1969). This work extending over eight years or more we may liken to the exhausting study of barnacles by Darwin in its care in observation and comparison. And as all good taxonomic work, it is the necessary basis for all of the practical applications of the biological knowledge to problems of ecology and agriculture.

Finally Darlington wrote (p. 363),

"In conclusion, I want to come back to what I think is *the* primary characteristic of good modern taxonomy—increase in the reality of classifications, reached by continual comparison with nature. . . . Taxonomy began among primitive peoples who distinguished and gave local names to the real, useful or dangerous (and presumably also the beautiful or curiosity-rousing) species of plants and animals around them. Many of the trends of taxonomy since then, from typology to some mathematical taxonomy, have been away from reality. But now

we are coming back—at least I think and hope we are—to a taxonomy
that is correlated with reality and that is becoming increasingly useful
as a base for diverse biological studies of living plants and animals"
(p. 363).

Some principles of practical taxonomy

1. Any organism can be known directly only from its somatic character
expressions, that is its phenotype.
2. Organisms live and die by these phenotypes, for natural selection
can act directly only upon expressed characters. Such characters may
be correlated with hidden characters, which therefore may be acted
upon in an indirect manner, but the primary action is always directly
upon the phenotype.
3. The genetic basis for hereditary characters is identified by, inferred
from, and naturally selected for or against, through particular somatic
character expressions. There is no knowledge of genetic mechanisms
other than through knowledge of somatic characters.
4. The whole office of the genotype is to produce and reproduce viable
phenotypes, and there is no other biological or taxonomic value
resident in such genotypes. Natural selection will pass any genotype
which can produce viable and adaptable phenotypes. Consequently,
the theory that knowledge of genotypes is always superior to knowledge
of phenotypes as a basis for taxonomic arrangements is without
scientific justification.
5. Natural selection has obviously passed many lines of convergent
evolution, and thus the practise of ignoring all convergence in taxonomic arrangements is un-natural and even, in part, non-evolutionary! It
is generally agreed that four-chambered hearts in birds and mammals
are a result of convergent evolution. But phylogenists would be horrified
at the thought that such similarities should be included in any taxonomic arrangements. This in spite of the fact that four-chambers and the
related warm-blooded metabolisms are of great importance to the
organisms affecting all their structures and activities. A vast amount of
predictivity resides in such characters, and it is wholly unwarranted to
ignore them in systematic appraisals.
6. We have been told without phylogeny, classification is meaningless

(Delevoryas, 1964) and that the "backward ideology" underlying phenetics should be rejected (Mayr, 1965). On the contrary we agree with Crowson (1910, p. 11) that classifying things (and ideas) is probably the most fundamental activity of the human mind and is basic to all forms of science. We say that systematic thinking is the first essential in systematic zoology. A new mysticism has developed regarding the genetic and phylogenetic bases for classification to which some of the statements of Hennig, Kiriakoff and Mayr bear witness. That Mayr should elevate "genetic programs" above the natures of the organisms themselves is a contradiction in terms and reveals a lack of judgment in these particulars on the part of one who is generally sound. For example Mayr said (1965, p. 98) "He who ignores that all organisms are united by descent throws away a good deal of available information". The fact is there is no such information, available or otherwise, and we deplore the practise of attempting to pass off mere hypotheses as knowledge, a practise characteristic of so many phylogenetic taxonomists. Even if it could be known that all organisms have genetic relationship, this knowledge could not account for the great diversities among organisms. It is perfectly possible to do evolutionary taxonomy as Darwin did with his barnacles, without so much unwarranted speculation.

7. There is no way to measure genetic relationships so as to be able to translate such values into classifications, and no biological justification for the attempt to do so. We believe Simpson (1961) was quite correct in the following remarks.

"What matters in taxonomy, and indeed in evolution as a whole, is the working organism as such and not primarily its genes. If a given characteristic is continuously present in an ancestor and in all the descendants of a given lineage, then it is homologous throughout even though the genetic substrate has changed. Of course this does not alter the fact that evolution of all characters does always have a genetic substrate. It only points out that genetic evolution and somatic evolution are not identical or precisely parallel and that it is somatic evolution that is more directly pertinent in taxonomy."*

Neither the number of like genes, evident within the limits of inter-

* From *Principles of Animal Taxonomy* by G. G. Simpson, Columbia University Press (1961). With permission.

crossing, nor their proportion to total numbers, nor the proportions of conjugating chromatin revealed by the experiments of Bolton and McCarthy (1962) have any simple or constant relationship to the amounts of similarity in somatic character expressions, that is the phenotypes.

These phenotypes constitute the natures of the organisms and a truly natural classification must have these as its basis. Relative weighting among somatic characters in accordance with their consistency and patterning is to be expected and that taxonomists will vary in such weighting is also to be expected. But the *a priori* dictum that ancestral characters are always better than descendent characters for the classification of descendent organisms should be abandoned. In a sense it may be said that there is nothing so backward in taxonomic theory as the insistence on supposed ancestry or "propinquity of descent" as the sole basis for classifying organisms, existing or fossil.

We have had more than a century of the backward look in systematics, following Darwin's *Origin of Species*. It is unfortunate that both Darwin's practise as clearly shown in his Monographs on the Cirripedia and even his theory as presented in the *Origin* have been so largely ignored, misrepresented, or misunderstood. He showed how it was possible to be an evolutionary taxonomist without submission to unwarranted and unnecessary speculation.

Darlington (1971) has pointed out, as we indicated, that the theory of ancestry-phylogeny has come to occupy the attention of a great many taxonomists who publish in systematic zoology. Even the computers, which could better be used in appraising overall similarities among organisms, have been diverted to the study of branching sequences, among real or imaginary organisms (Camin and Sokal, 1965). This smacks more of "fun and games" than the practical business of taxonomy. We recommend that the attention of systematists be focused upon organisms as they are and as natural selection must deal with them. Fossils and existing organisms should be described, compared and grouped on the basis of what they are without making unverifiable claims as to the remote or even the recent course of evolution and their phylogenetic history. Darwin constructed a lasting classification of barnacles on such an objective basis. We can only hope that evolutionary taxonomists will more largely follow his example.

We are indebted to Robert R. Sokal (1972) for calling our attention to the recent book *Mathematical Taxonomy* by Jardine and Sibson (1971). Under the heading "Taxonomy Rationalized" Sokal briefly reviews this book which is divided into three main sections. These deal with (a) dissimilarity measures, (b) clustering methods and (c) the principles of mathematical and biological taxonomy. We have no competence in the field of mathematical taxonomy and must leave the contributions of Jardine and Sibson for Sokal and others to evaluate, but we fully agree that the third section "is a gem". Sokal believes that Jardine and Sibson belong to a new generation, who are able to deal with the problems of modern biological taxonomy in a remarkably clear and objective way, and without being influenced by the many emotional hang-ups of previous taxonomists. With this appraisal we agree completely, but find it impossible to paraphrase the discussions of Jardine and Sibson relating to (a) the demarcation of taxa of specific rank, (b) the grouping of taxa of specific rank into taxa of supraspecific rank, (c) the grouping of populations within taxa of specific rank into taxa of infraspecific rank.

Biological species definitions, cladism, pheneticism, phylogenetic weighting, and all the other problems of modern taxonomy are dealt within an even-handed way but in such condensed writing that we cannot abstract them for inclusion here. What we do recommend is that taxonomists of all kinds, numerically inclined or not, study carefully the third part of the book which can be understood by all.

We support the general views regarding systematics of Blackwelder (1962, 1964, 1967), Cain (1959 a and b, 1963), Gilmour (1940, 1961), Heywood (1964) and others, who would recapture for biologists a practical and meaningful taxonomy, which though evolutionary of necessity, yet reduces speculation as much as possible. In this connection Heywood (1964) wrote about this coming *taxonomic* revolution, as a *realist* phase of taxonomy. In it there must be an acknowledgement that the most generally useful classification results from group-making based on overall resemblances. This is a phenetic approach which involves comparison of many of the available characters from which a selection is made *a posteriori* on the basis of their availability and consistency, never on the basis of the *a priori* notion that a phylogenetic approach is necessarily superior to any other for a natural classification. "Let honesty and realism be the keynote of the taxonomic revolution" (*op. cit.*, p. 47). And above all let Darwin's Monographs be the classical examples of this honest and practical taxonomy.

REFERENCES

AKABORI, S. (1960) On the origin of the fore-protein. In Florkin, M., editor, *Aspects of the Origin of Life*. Pergamon Press, London.

ALLEN, J. M., editor (1963) *The Nature of Biological Diversity*. McGraw-Hill, New York.

ASCOLI, M. (1902) Ueber den Mechanismus der Albuminurie durch Eiereiweis. *Munch. Med. Wochensch.* **49**, 398.

BAER, K. E. VON (1828) Ueber Entwickelungsgeschichte der Thiere. Königsberg Impression Anastaltique (1967). *Culture et Civilisation*, Bruxelles.

BAIER, J. G. (1959) Light-transmission measurements in systematic serology using the "Spectronic 20". *Bull. Serol. Mus.* No. **21**, 5.

BAKER, J. R. (1948) The status of the Protozoa. I. *Nature*, London, **161**, 548.

BARGHOORN, E. S. (1971) The oldest fossils. *Sci. Amer.* **224**, 30.

BARGHOORN, E. S. and SCHOPF, J. W. (1966) Microorganisms three billion years old from the Pre-Cambrian of South Africa. *Science*, **152**, 758.

BARGHOORN, E. S. and TYLER, S. A. (1965) Microorganisms from the Gunflint chert. *Science*, **147**, 563.

BASFORD, N., BUTLER, J. E., LEONE, C. A. and ROHLF, F. J. (1968) Immunological comparisons of selected Coleoptera with analyses of relationships using numerical taxonomic methods. *Syst. Zool.* **17**, 388.

BATHER, F. A. (1927) Biological classification, past and future. *Quart. J. Geol. Soc.* **83**, 62.

BAUM, W. C. (1954) Systematic serology of the family Cucurbitaceae, with special reference to the genus *Cucurbita*. *Bull. Serol. Mus.* No. **13**, 5.

BEADLE, G. W. (1946) The gene. *Proc. Amer. Philos. Soc.* **90**, 422.

BEADLE, G. W. and COONRADT, V. (1944) Heterocaryosis in *Neurospora crassa*. *Genetics* **29**, 291.

BERKNER, L. V. and MARSHALL, L. C. (1965) On the origin and rise of oxygen concentration in the Earth's atmosphere. *Jour. Atmos. Sci.* **22**, 225.

BERNAL, J. D. (1951) *The Physical Basis of Life*. Routledge and Kegan Paul, London.

BLACKWELDER, R. E. (1962) Animal taxonomy and the new systematics. In *Survey of Biological Progress*. Vol. **IV**, Acad. Press, New York.

BLACKWELDER, R. E. (1964) Phyletic and phenetic versus omnispective classification. In *Phenetic and Phylogenetic Classification*, edited by Heywood, V. H. and McNeill, J., The Systematics Assn., London.

BLACKWELDER, R. (1967) *Taxonomy—A Text and Reference Book*. John Wiley & Sons, New York.

BLACKWELDER, R. E. and BOYDEN, A. (1952) The nature of systematics. *Syst. Zool.* **1**, 26.

BLYTH, E. (1835) An attempt to classify the varieties of animals with observations on the marked seasonal and other changes which take place in various British species and which do not constitute varieties. *Ann. and Mag. Natur. Hist.* **8**, 40.

BOCK, W. J. (1963) Evolution and phylogeny in morphologically uniform groups. *Amer. Natur.* **97**, 265.

BOLTON, E. T., LEONE, C. A. and BOYDEN, A. (1948) A critical analysis of the performance of the photronreflectometer in the measurement of turbid systems. *J. Immunol.* **58**, 169.

BOLTON, E. T. and MCCARTHY, B. J. (1962) A general method for the isolation of RNA complementary to DNA. *Proc. Nat. Acad. Sci.* **48**, 1390.

BONNER, J. T. (1958) *The Evolution of Development.* Cambridge University Press.

BOOR, A. K. and HEKTOEN, L. (1930) Preparation and properties of carbon monoxide hemoglobin. *J. Infec. Dis.* **46**, 1.

BORDET, J. (1899) Sur l'agglutination et la dissolution des globules rouges par le serum d'animaux injectes de sang defibriné. *Ann. de l'Inst. Pasteur* **13**, 688.

BOYD, W. C. (1950) *Genetics and the Races of Man.* Little, Brown & Co., Boston.

BOYD, W. C. (1956) *Fundamentals of Immunology*, 3rd ed. Interscience Publishers, Inc., New York.

BOYD, W. C. (1962) *Introduction to Immunochemical Specificity.* John Wiley & Sons, Inc., New York.

BOYDEN, A. (1926) The precipitin reaction in the study of animal relationships. *Biol. Bull.* **50**, 73.

BOYDEN, A. (1932) Precipitin tests as a basis for a quantitative phylogeny. *Proc. Soc. Exper. Biol. & Med.* **29**, 955.

BOYDEN, A. (1934) Precipitins and phylogeny in animals. *Amer. Natur.* **68**, 516.

BOYDEN, A. (1935) Genetics and homology. *Quart. Rev. Biol.* **10**, 448.

BOYDEN, A. (1939) Serological study of the relationships of some common Invertebrata. *Carnegie Inst. Wash. Yearbook* No. **38**, 219.

BOYDEN, A. (1942) Systematic Serology: a critical appreciation. *Physiol. Zool.* **15**, 109.

BOYDEN, A. (1943a) Homology and analogy. A century after the definitions of "Homologue" and "Analogue" of Richard Owen. *Quart. Rev. Biol.* **18**, 228.

BOYDEN, A. (1943b) Serology and animal systematics. *Amer. Natur.* **77**, 234.

BOYDEN, A. (1947) Homology and analogy. A critical review of the meanings and implications of these concepts in biology. *Amer. Mid. Natur.* **37**, 648.

BOYDEN, A. (1948a) The Serological Museum. *Bull. Serol. Mus.* No. **1**, 1.

BOYDEN, A. (1948b) The Serological Museum. *The Biologist* **31**, 10.

BOYDEN, A. (1948c) "Sereology" and "Immunology". *Science* **108**, 635.

BOYDEN, A. (1949) The Serological Museum, Bureau of Biological Research. Rutgers University. *Turtox News* **27**, 204.

BOYDEN, A. (1950) Is parthenogenesis sexual or asexual reproduction? *Nature, Lond.* **166**, 820.

BOYDEN, A. (1951) A half-century of systematic serology: The work of Nuttall at Cambridge and of his successors in America. *Bull. Serol. Mus.* No. **6**, 1.

BOYDEN, A. (1953a) Comparative evolution with special reference to primitive mechanisms. *Evolution* **7**, 21.

BOYDEN, A. (1953b) Fifty years of systematic serology. *Syst. Zool.* **2**, 19.

BOYDEN, A. (1954a) The significance of asexual reproduction. *Syst. Zool.* **3**, 26.

BOYDEN, A. (1954b) The measurement and significance of serological correspondence among proteins. In Cole, W. H., editor. *Serological Approaches to Studies of Protein Structure and Metabolism.* Rutgers University Press, New Brunswick.

BOYDEN, A. (1957a) Are there any "acellular animals"? *Science* **125**, 155.

BOYDEN, A. (1957b) Concerning the "cellularity" or acellularity of the protozoa. *Science* **125**, 990.

BOYDEN, A. (1958) Comparative serology: aims, methods and results. In Cole, W. H., editor. *Serological and Biochemical Comparisons of Proteins*. Rutgers University Press, New Brunswick, N.J.

BOYDEN, A. (1959a) Serology as an aid to systematics. *Bull. Serol. Mus.* No. **22**, 4.

BOYDEN, A. (1959b) Serology as an aid to systematics. *Proc. XVth Internat. Cong. Zoology*, London.

BOYDEN, A. (1960) A brief commentary on Simpson's anatomy and morphology: classification and evolution 1859 and 1951. *Syst. Zool.* **9**, 44.

BOYDEN, A. (1964) Perspectives in systematic serology. In Leone, C. A., editor. *Taxonomic Biochemistry and Serology*. Ronald Press Co., New York.

BOYDEN, A. (1969) Homology and analogy. *Science* **164**, 455.

BOYDEN, A. (1970) A critique of some recent reports in the field of systematic serology. *Bull. Serol. Mus.* No. **44**, 8.

BOYDEN, A. (1971) Concerning the specificity of precipitin and other serological reactions. *Bull. Serol. Mus.* No. **46**, 1.

BOYDEN, A. and NOBLE, G. K. (1933) The relationships of some common Amphibia as determined by serological study. *Amer. Mus. Novitates* No. **606**, 1.

BOYDEN, A. and DE FALCO, R. J. (1943) Report on the use of the photronreflectometer in serological comparisons. *Physiol. Zool.* **16**, 229.

BOYDEN, A., BOLTON, E. and GEMEROY, D. (1947) Precipitin testing with special reference to the photoelectric measurement of turbidity. *J. Immunol.* **57**, 211.

BOYDEN, A., LEONE, C. and BOLTON, E. (1949) The Serological Museum of Rutgers University. *13th Internat. Cong. of Zoology*, Paris, 1948.

BOYDEN, A., DE FALCO, R. J. and GEMEROY, D. G. (1950a) A New Principle. *Bull. Serol. Mus.* No. **4**, 3.

BOYDEN, A. and GEMEROY, D. (1950b) The relative position of the Cetacea among the orders of Mammalia as indicated by precipitin tests. *Zoologica* **35**, 145.

BOYDEN, A., DE FALCO, R. J. and GEMEROY, D. G. (1951) Parallelism in serological correspondence. *Bull. Serol. Mus.* No. **6**, 3.

BOYDEN, A., GEMEROY, D. and DE FALCO, R. (1956) On measuring serological correspondence among antigens. *Bull. Serol. Mus.* No. **16**, 3.

BOYDEN, Alan and PAULSEN, Elizabeth C. (1957) Significance of electrophoretic patterns in systematic serology. *Bull. Serol. Mus.* No. **18**, 7.

BOYDEN, A. and SHELSWELL, E. M. (1959) Prophylogeny: some considerations regarding primitive evolution in lower Metazoa. *Acta Biotheor.* **13**, 115.

BOYDEN, A. and PAULSEN, E. C. (1959b) Serology as an aid to systematics. *Bull. Serol. Mus.* No. **22**, 4.

BREDER, C. M., JR. (1942) A consideration of evolutionary hypotheses in regard to the origin of life. *Zoologica* **27**, 131.

BREUER, M. E. and PAVAN, C. (1955) Behavior of polytenechromosomes of *Rhynchosciara angelae* at different stages of larval development. *Chromosoma* **7**, 371.

BRIGGS, R. and KING, T. J. (1956) Serial transplantation of embryonic nuclei. *Cold Spring Harbor Symp. Quant. Biol.* **21**, 271.

BRIGGS, R., SIGNORET, J. and HUMPHREY, R. R. (1964) Transplantation of nuclei of various cell types from neurulae of the Mexican axolotl. (*Ambystoma mexicanum*) *Develop. Biol.* **10**, 233.

BRITTEN, R. J. (1969) Repeated DNA and transcription. In *Problems in Biology: RNA in Development*, Hanly, E. W., editor. University of Utah Press, Salt Lake City.

BRITTEN, R. J. and KOHNE, D. E. (1968) Repeated sequences in DNA. *Science* **161**, 529.

BUTLER, J. E. and LEONE, C. A. (1968) Determination of immunologic correspondence for taxonomic studies by densitometric scanning of antigen-antibody precipitates in agar-gel. *Comp. Biochem. Physiol.* **25**, 417.

CAIN, A. J. (1959a) Deductive and inductive methods in post-Linnaean taxonomy. *Proc. Linn. Soc. Lond.* **170**, Session 185.

CAIN, A. J. (1959b) The post-Linnaean development of taxonomy. *Proc. Linn. Soc. Lond.* **170**, Session 234.

CAIN, A. J. (1963) The perfection of animals. *Viewpoints in Biol.* **3**, 36.

CAMIN, J. H. and SOKAL, R. R. (1965) A method for deducing branching sequences in phylogeny. *Evolution* **19**, 311.

CAWLEY, L. P. (1969) *Electrophoresis and Immunoelectrophoresis.* Little, Brown & Co., Boston.

CEI, J. (1963) Some precipitin tests and remarks on the systematic relationships of four South American families of frogs. *Bull. Serol. Mus.* No. **30**, 4.

CEI, J. (1965) The relationships of some Ceratophryid and Leptodactylid genera as indicated by precipitin tests. *Herpetologica* **20**, 217.

CLARK, A. H. (1930) *The New Evolution.* Williams and Wilkins, Baltimore.

CLEVELAND, L. R. (1949) Hormone-induced sexual cycles of Flagellates. I. Gametogenesis, fertilization, and meiosis in *Trichonympha. J. Morph.* **85**, 197.

CLEVELAND, L. R. (1950) Hormone-induced sexual cycles of Flagellates. II. Gametogenesis, fertilization, and one-division meiosis in *Oxymonas. J. Morph.* **86**, 185. III. Gametogenesis, fertilization, and one-division meiosis in *Saccinobaculus. J. Morph.* **86**, 215. IV. Meiosis after syngamy and before nuclear fusion in *Notila. J. Morph.* **87**, 317. V. Fertilization in *Euconympha. J. Morph.* **87**, 349.

CLEVELAND, L. R. (1951) Hormone-induced sexual cycles of Flagellates. VI. Gametogenesis, fertilization, meiosis, oöcysts and gametocysts in *Leptospironympha. J. Morph.* **88**, 199. VII. One-division meiosis and autogamy without cell division in *Urinympha. J. Morph.* **88**, 385.

COHEN, E. (1955) Immunological studies of the serum proteins of some reptiles. *Biol. Bull.* **109**, 394.

COLE, W. H., editor (1954) *Serological Approaches to Studies of Protein Structure and Metabolism.* Rutgers University Press, New Brunswick, N.J.

COLE, W. H., editor (1958) *Serological and Biochemical Comparisons of Proteins.* Rutgers University Press, New Brunswick, New Jersey.

COMMONER, B. (1962) Is DNA a Self-duplicating Molecule? *Horizons in Biochemistry.* Acad. Press, N.Y.

COMMONER, B. (1968) Failure of the Watson–Crick theory as a chemical explanation of inheritance. *Nature* **220**, 334.

CORLISS, J. O. (1957) Concerning the "cellularity" or acellularity of the Protozoa. *Science* **125**, 988.

CORLISS, J. O. (1959) Comments on the systematics and phylogeny of the Protozoa. *Syst. Zool.* **8,** 169.

CRACRAFT, J. (1967) Comments on homology and analogy. *Syst. Zool.* **16,** 356.

CREW, F. A. E. (1925) *Animal Genetics: An Introduction to the Science of Animal Breeding.* Oliver & Boyd, Edinburgh.

CRICK, F. H. C. (1958) On protein synthesis in biological replication of macromolecules. *Soc. Exp. Biol. Symp.* **12,** 138.

CROSBY, J. L. (1955) The evolution of mitosis. *Proc. Univ. Durham Phil. Soc.* **12,** 73.

CROWLE, A. J. (1961) *Immunodiffusion.* Academic Press, New York.

CROWSON, R. A. (1970) *Classification and Biology.* Aldine-Atherton Press, Inc., Chicago.

CUMLEY, R. W. and IRWIN, M. R. (1942) Immunogenetic studies of species: segregation of serum components in backcross individuals. *Genetics* **27,** 177.

CUMLEY, R. W. and IRWIN, M. R. (1940) Differentiation of sera of two species of doves and their hybrid. *Proc. Soc. Exp. Biol. Med.* **44,** 353.

CUVIER, G. (1817) *Le Regne Animal distribué d'apres son Organisation pour servir de base a l'Histoire Naturelle des Animaux, et d'Introduction a l'Anatomie Comparee* Paris.

DANIELLI, J. F. (1959) Some theoretical aspects of nucleocytoplasmic relationships. *Exp. Cell. Res.* Supp. **6,** 252.

DANIELLI, J. F., LORCH, I. J., ORD, M. J. and WILSON, E. G. (1955) Nucleus and cytoplasm in cellular inheritance. *Nature,* Lond. **176,** 1114.

DARLINGTON, C. D. (1958) *Evolution of Genetic Systems.* 2nd edition. Oliver and Boyd, Edinburgh.

DARLINGTON, C. D. (1959) *Darwin's Place in History.* Blackwell, Oxford.

DARLINGTON, P. J., JR. (1971) Modern taxonomy, reality, and usefulness. *Syst. Zool.* **20,** 341.

DARWIN, C. (1851) *A Monograph on the sub-class Cirripedia.* The Lepadidae; or, Pedunculated Cirripedes. London, Printed for the Ray Society.

DARWIN, C. (1854) *A Monograph on the sub-class Cirripedia. The Balanidae, The Verrucidae,* etc. London, Printed for the Ray Society.

DARWIN, C. (1859) *On the Origin of Species.* John Murray, London.

DARWIN, C. (1868) *The Variation of Animals and Plants under Domestication.* Orange Judd & Co., New York.

DARWIN, C. (1882) *The Formation of Vegetable Mould, through the Action of Worms, with Observations on Their Habits.* D. Appleton & Co., New York.

DARWIN, F., editor (1903) *More Letters of Charles Darwin.* John Murray, London.

DARWIN, F., editor (1909) *The Foundations of the Origin of Species.* Cambridge University Press.

DAWES, B. (1952) *A Hundred Years of Biology.* Duckworth, London.

DE BEER, G. R. (1928) *Vertebrate Zoology.* Macmillan, New York.

DE BEER, G. R. (1958) *Embryos and Ancestors,* 3rd edition. Oxford University Press.

DE BEER, G. R. (1959) Paedomorphosis. *Proc. 15th Int. Congr. of Zool.,* 927.

DE FALCO, R. J. (1942) A serological study of some Avian relationships. *Biol. Bull.* **83,** 205.

DELEVORYAS, T. (1964) The role of palaeobotany in vascular plant classification. In Heywood, V. H. and McNeill, J., editors. *Phenetic and Phylogenetic Classification.* Published by The Systematics Association, London.

K

DESSAUER, H. C. and FOX, W. (1964) Electrophoresis in taxonomic studies, illustrated by analyses of blood proteins. In Leone, C. A., editor, *Taxonomic Biochemistry and Serology.* Ronald Press Co., New York.

DE VRIES, H. (1910) *Intracellular Pangenesis.* Open Court Pub. Co., Chicago.

DILLON, L. S. (1962) Comparative cytology and the evolution of life. *Evolution* **16,** 102.

DILLON, L. S. (1963) A reclassification of the major groups of organisms based upon comparative cytology. *Syst. Zool.* **12,** 71.

DOBELL, C. C. (1911) The principles of protistology. *Arch. f. Protistenk.* **23,** 269.

DOBZHANSKY, Th. (1937) *Genetics and the Origin of Species.* Columbia University Press.

DOBZHANSKY, T. (1959) Blyth, Darwin and natural selection. *Amer. Natur.* **93,** 204.

DOUGHERTY, E. (1955) The origin of sexuality. *Syst. Zool.* **4,** 145.

DOUGHERTY, E. C. and ALLEN, M. B. (1960) Is pigmentation a clue to protistan phylogeny? In Allen, M. B., editor, *Comparative Biochemistry of Photoreactive Systems.* Acad. Press, New York.

DOUGHERTY, E. C., editor (1963) *The Lower Metazoa; Comparative Biology and Phylogeny.* Univ. Calif. Press, Berkeley and Los Angeles.

DUERDEN, J. E. (1923–1924) Methods of evolution. *Science Prog.* **18,** 556.

DUNSFORD, I. and BOWLEY, C. C., 2nd edition (1967) *Techniques in Blood Grouping.* Charles C. Thomas. Springfield, Ill.

EHRENSVÄRD, G. (1962) *Life: Origin and Development.* Chicago University Press.

EISELEY, L. C. (1959) Charles Darwin, Edward Blyth, and the theory of natural selection. *Proc. Amer. Phil. Soc.* **103,** 94.

EISENBRANDT, L. L. (1938) On the serological relationship of some helminths. *Amer. J. Hyg.* **27,** 117.

EHRHARDT, A. (1929) Der Wert der Immunitätsreaktionen fur Phylogenetische Untersuchungen in der Zoologie. *Zeit. f. Immunitätsf.* **60,** 156.

EHRHARDT, A. (1930) Die serdiagnostischen Untersuchungen in der Ornithologie. *J. f. Ornith.* **78,** 214.

EHRHARDT, A. (1931) Die Verwandtschaftsbestimmungen mittels der Immunitätsreaktionen in der Zoologie und ihr Wert für phylogenetische Untersuchungen. *Ergeb. u. Fortsch. Zool.* **7,** 279.

FAIRBROTHERS, D. E. (1966) Comparative serological studies in plant systematics. *Bull. Serol. Mus.* No. **35,** 2.

FAIRBROTHERS, D. E. (1969) Plant serotaxonomic (serosystematic) literature 1951–1968. *Bull. Serol. Mus.* No. **41,** 1.

FAIRBROTHERS, D. E. and JOHNSON, M. A. (1961) The precipitin reaction as an indicator of the relationship in some grasses in *Recent Advances in Botany.* University Toronto Press, Vol. **1,** 116.

FESENKOV, V. G. (1960) Some considerations on the primaeval state of the earth. In Florkin, M., editor. *Aspects of the Origin of Life.* Pergamon Press, Lond.

FLORKIN, M., editor (1960) *Aspects of the Origin of Life.* Pergamon Press, New York.

FLOWER, W. H. and LYDDEKER, R. (1891) *An Introduction to the Study of Mammals Living and Extinct.* Adam and Charles Black, Lond.

FOX, S. W. (1960) A chemical theory of spontaneous generation. In Florkin, M., editor. *Aspects of the Origin of Life.* Pergamon Press, Lond.

FRAIR, W. (1964) Turtle family relationships as determined by serological tests. In Leone, C. A., editor, *Taxonomic Biochemistry and Serology.* The Ronald Press Co., New York.

FRAIR, W. (1969) Aging of serum proteins and serology of marine turtles. *Bull. Serol. Mus.* No. **42,** 1.

FRASER, R. C. (1959) Cytodifferentiation; protein synthesis in transition. *Amer. Natur.* **93,** 47.

FREDERICK, J. F. (1961) Immunochemical studies of phosphorylases of Cyanophyceae. *Phyton.* **16,** 21.

GABRIEL, M. L. (1960) Primitive genetic mechanisms and the origin of chromosomes. *Amer. Natur.* **94,** 257.

GAFFRON, H. (1960) The origin of life. In Tax, S., editor, *Evolution after Darwin.* University of Chicago Press.

GARSTANG, W. (1922) The theory of recapitulation: a critical restatement of the biogenetic law. *J. Linn. Soc. Lond.* **35,** 81.

GEGENBAUR, C. (1878) *Elements of Comparative Anatomy.* Trans. by F. J. Bell. Macmillan, London.

GELL, P. G. H., HAWKES, J. G. and WRIGHT, S. T. C. (1960) The application of immunological methods to the taxonomy of species within the genus *Solanum. Proc. Roy. Soc. Lond.,* Series B. **151,** 364.

GEMEROY, D. G. (1943) On the relationship of some common fishes as determined by the precipitin reaction. *Zoologica* **28,** 109.

GEMEROY, D. (1952) Identification of Primate sera. *Bull. Serol. Mus.* No. **9,** 7.

GEMEROY, D., BOYDEN, A. and DE FALCO, R. (1955) What blood is that? *Bull. Serol. Mus.* No. **14,** 6.

GEMEROY, D. and BOYDEN, A. (1961) Preliminary report on precipitin tests with American and European eel sera. *Bull. Serol. Mus.* No. **26,** 7.

GHISELIN, M. T. (1969) *The Triumph of the Darwinian Method.* University of California Press, Berkeley.

GILMOUR, J. S. L. (1940) Taxonomy and Philosophy. In Huxley, J., editor, *The New Systematics.* Oxford University Press, London, New York.

GILMOUR, J. S. L. (1951) The development of taxonomic theory since 1851. *Nature,* Lond. **168,** 400.

GILMOUR, J. S. L. (1961) Taxonomy. In MacLeod, A. M. and Cobley, L. S., editors, *Contemporary Botanical Thought.* Oliver and Boyd, Edinburgh.

GLAESSNER, M. F. (1961) Pre-Cambrian animals. *Sci. Amer.* **204,** 72.

GLAESSNER, M. F. (1962) Pre-Cambrian fossils. *Biol. Rev.* **37,** 467.

GLAESSNER, M. F. and WADE, M. (1966) The Late Pre-Cambrian fossils from Ediacara, South Australia. *Palaeontol.* **9,** 599.

GLENN, W. G. (1957) Direct photometry of diffusing precipitin systems for characterizing proteins. *Bull. Serol. Mus.* No. **18,** 1.

GLENN, W. G. (1958) Characterization of precipitin systems by direct photometry of agar columns. In Cole, W. H., editor, *Serological and Biochemical Comparisons of Proteins.* Rutgers University Press, New Brunswick, N.J.

GOLDSCHMIDT, R. B. (1940) *The Material Basis of Evolution.* Yale University Press, New Haven.

GOODMAN, M. (1960a) On the emergence of intraspecific differences in the protein antigens of human beings. *Amer. Natur.* **94,** 153.

GOODMAN, M. (1960b) The species specificity of proteins as observed in the Wilson comparative analyses plates. *Amer. Natur.* **94,** 184.

GOODMAN, M. (1961) The role of immunochemical differences in the phyletic development of human behavior. *Human Biol.* **33,** 131.

GOODMAN, M. (1963) Man's place in the phylogeny of the Primates as reflected in serum proteins. In Washburn, S. L., editor, *Classification and Human Evolution.* Aldine Pub. Co., Chicago.

GOODMAN, M. (1964) Problems of Primate systematics attacked by the serological study of proteins. In Leone, C. A., editor, *Taxonomic Biochemistry and Serology.* Ronald Press Co., New York.

GOODMAN, M. and MOORE, G. W. (1971) Immunodiffusion systematics of the Primates. I. The Catarrhini. *Syst. Zool.* **20,** 19.

GRABAR, P. and WILLIAMS C. A., JR. (1953) Methode permettant l'etude conjugée des proprietés electophoretiques et immunochimique d'un melange des proteines. *Biochim. Biophys. Acta* **10,** 193.

GREENBERG, M. J. (1959) Ancestors, embryos, and symmetry. *Syst. Zool.* **8,** 212.

GREGG, J. R. (1959) On deciding whether protistans are cells. *Phil. of Sci.* **26,** 338.

GREGG, J. R. and HARRIS, F. T. C. (1964) *Form and Strategy in Science.* D. Reidel Pub. Co., Dordrecht, Holland.

GRIMSTONE, A. V. (1959) Cytology, homology and phylogeny—a note on "Organic Design". *Amer. Natur.* **93,** 273.

GURDON, J. B. (1962) The developmental capacity of nuclei taken from intestinal epithelium cells of feeding tadpoles. *J. Embryol. Exp. Morph.* **10,** 622.

GURDON, J. B. (1963) Nuclear transplantation in Amphibia and the importance of stable nuclear changes in promoting cellular differentiation. *Quart. Rev. Biol.* **38,** 54.

GURDON, J. B. (1964) The transplantation of living cell nuclei. *Advance Morphogen.* **4,** 1.

GURDON, J. B. (1967) Control of gene activity during the early development of *Xenopus laevis.* In Brink, R. A., editor, *Heritage from Mendel.* University of Wisconsin Press, Madison.

GURDON, J. B. (1969) The importance of egg cytoplasm for the control of RNA and DNA synthesis in early Amphibian development. In Hanly, E. W., editor. *Problems in Biology: RNA in Development.* University of Utah Press, Salt Lake City.

GURDON, J. B. and UEHLINGER, V. (1966) "Fertile" intestine nuclei. *Nature* **210,** 1240.

HAAS, O. and SIMPSON, G. G. (1946) Analysis of some phylogenetic terms with attempts at redefinition. *Proc. Amer. Phil. Soc.* **90,** 319.

HADŽI, J. (1953) An attempt to reconstruct the system of animal classification. *Syst. Zool.* **2,** 145.

HADŽI, J. (1963) *The Evolution of the Metazoa.* Pergamon Press, Macmillan Co., New York.

HAECKEL, E. (1866) *Generelle Morphologie der Organismen.* G. Reiner, Berlin.

HAFLEIGH, A. S. and WILLIAMS, C. A., JR. (1966) Antigenic correspondence of serum albumins among the Primates. *Science* **151,** 1530.

HALDANE, J. B. S. (1932) The Origin of Life. In *The Inequality of Man and Other Essays.* Chatto & Windus, London.

HALDANE, J. B. S. (1950) Some alternatives to sex. *New Biol.* **19,** 7.

HALDANE, J. B. S. (1954) *The Biochemistry of Genetics.* Allen & Unwin, Ltd., London.

HALL, T. S. (1951) *A Source Book in Animal Biology.* McGraw-Hill, New York.

HAMMOND, H. D. (1955a) Systematic serological studies in Ranunculaceae. *Bull. Serol. Mus.* No. **14**, 1.

HAMMOND, H. D. (1955b) A study of taxonomic relationship within the Solanaceae as revealed by the photroner serological method. *Bull. Serol. Mus.* No. **14**, 3.

HAND, C. (1959) On the origin and phylogeny of the Coelenterates. *Syst. Zool.* **8**, 191.

HAND, C. (1963) The early worm: a planula. In Dougherty, E. C., editor, *The Lower Metazoa.* University of California Press, Berkeley.

HANSON, E. D. (1958) On the origin of the Eumetazoa. *Syst. Zool.* **7**, 16.

HARTMANN, M. (1956) *Die Sexualität.* 2nd edition. Fischer, Stuttgart.

HAWES, R. S. J. (1963) The emergence of asexuality in Protozoa. *Quart. Rev. Biol.* **38**, 234.

HEIDELBERGER, M. (1954) The precipitin reaction and studies of native and denatured proteins and derivatives. In Cole, W. H., editor, *Serological Approaches to Studies of Protein Structure and Metabolism.* Rutgers University Press, New Brunswick, New Jersey.

HEIDELBERGER, M. and KENDALL, F. E. (1929) A quantitative study of the precipitin reaction between type III pneumococcus polysaccharide, and purified homologous antibody. *J. Exper. Med.* **50**, 809.

HEIDELBERGER, M., KENDALL, F. E. and TEORELL, T. (1936) Quantitative studies on the precipitin reaction. Effect of salts on the reaction. *J. Exper. Med,* **63**, 819.

HEKTOEN, L. (1922) Precipitin reactions of the normal and cataracterous lens. *J. Infec. Dis.* **31**, 72.

HEKTOEN, L. and BOOR, A. K. (1931) The specificness of hemoglobin precipitins. *J. Infec. Dis.* **49**, 29.

HENFREY, A. and HUXLEY, T. H., editors (1853) *Scientific Memoirs, Natural History.* Taylor and Francis, London.

HENNIG, W. (1950) *Grundzüge einer Theorie der phylogenetischen Systematik.* Deutscher Zentralverlag, Berlin.

HENNIG, W. (1966) *Phylogenetic Systematics.* University of Illinois Press, Urbana.

HEYWOOD, V. H. (1964) Introduction to General Principles. In Heywood, V. H. and McNeill, J., editors, *Phenetic and Phylogenetic Classification.* The Systematics Association, London.

HEYWOOD, V. H. and MCNEILL, J., editors (1964), *Phenetic and Phylogenetic Classification.* The Systematics Association, British Museum (Natural History), London.

HIS, W. (1888) On the principles of animal morphology. *Proc. Roy. Soc. Edinburgh* **15**, 294.

HOOKER, S. B. and BOYD, W. C. (1934) The existence of antigenic determinants of diverse specificity in a single protein. *J. Immunol.* **26**, 469.

HOOKER, S. B. and BOYD, W. C. (1936) The existence of antigenic determinants of diverse specificity in a single protein. III. Further notes on crystalline hen and duck ovalbumins. *J. Immunol.* **30**, 41.

HOOKER, S. B. and BOYD, W. C. (1941) Widened reactivity of antibody produced by prolonged immunization. *Proc. Soc. Exp. Biol. Med.* **47**, 187.

HOLMES, A. (1947) An estimate of the age of the earth. *Geol. Mag.,* London **84**, 123.

HOLMES, A. (1954) The oldest dated minerals of the Rhodesian shield. *Nature,* London, **173**, 612.

HOROWITZ, N. H. and STANLEY, L. MILLER (1962) Current theories on the origin of life. *Fortschr. der Chemie Organisch. Naturstoffe* **20**, 423.

HOYER, B. H., MCCARTHY, B. J. and BOLTON, E. T. (1964) A molecular approach in the systematics of higher organisms. *Science* **144**, 959.

HUBBS, C. L. (1944) Concepts of homology and analogy. *Amer. Natur.* **78**, 289.

HUTNER, S. H. and PROVASOLI, L. (1957) Concerning the "cellularity" or "acellularity" of the Protozoa. *Science* **125**, 989.

HUXLEY, J. (1928) In De Beer, G. R., *Vertebrate Zoology*. Macmillan, New York.

HUXLEY, J., editor (1940) *The New Systematics*. Oxford University Press, London.

HUXLEY, J., HARDY, A. and FORD, E. B. (1954) *Evolution as a Process*. Allen and Unwin, London.

HUXLEY, T. H. (1869) *An Introduction to the Classification of Animals*. John Churchill & Sons, London.

HUXLEY, T. H. (1894) Upon animal individuality. *Proc. Roy. Inst.* **1**, 184.

HYMAN, L. H. (1940) *The Invertebrates. Vol. 1. Protozoa through Ctenophora*. McGraw-Hill, New York.

HYMAN, L. H. (1942) The transition from the unicellular to the multicellular individual. In *Biol. Symp.* **8**, Jacques Cattell Press, Lancaster, Pa.

HYMAN, L. H. (1951) *The Invertebrates. Vol. 2. Platyhelminthes and Rhynchocoela. The Acoelomate Bilateria*. McGraw-Hill, New York.

HYMAN, L. H. (1959) *The Invertebrates. Vol. 5. Smaller Coelomate Groups*. McGraw-Hill, New York.

INGLIS, W. G. (1966) The observational basis of homology. *Syst. Zool.* **15**, 219.

JACOBSHAGEN, E. (1925) *Allgemeine vergleichende Formenlehre der Tiere*. W. Klinkhardt, Leipzig.

JÄGERSTEN, G. (1955) On the early phylogeny of the Metazoa. The bilaterogastrea theory. *Zool. Bidrag.* **30**, 321.

JÄGERSTEN, G. (1959) Further remarks on the early phylogeny of the Metazoa. *Zool. Bidrag. Fran Uppsala.* **33**, 79.

JARDINE, N. (1967) The concept of homology in biology. *Brit. J. Phil. Sci.* **18**, 125.

JARDINE, N. and SIBSON, R. (1971) *Mathematical Taxonomy*. John Wiley & Sons, Ltd., London.

JENNINGS, R. K. and KAPLAN, M. A. (1961) Qualitative comparative serology. *Bull. Serol. Mus.* No. **25**, 5.

JENSEN, U. (1968a) Serologische Beiträge zur Systematik der Ranunculaceae. *Bot. Jahrb.* **88**, 204.

JENSEN, U. (1968b) Serologische Beiträge zur Systematik der Ranunculaceae. *Bot. Jahrb.* **88**, 269.

JENSEN, U., FROHNE, D. and MORITZ, O. (1964) Serological investigations in the field of Rhoeadales and Ranunculaceae. *Bull. Serol. Mus.* No. **32**, 3.

JOHNSON, M. A. (1954) The precipitin reaction as an index of relationship in the Magnoliaceae. *Bull. Ser. Mus.* No. **13**, 1.

JOHNSON, M. and FAIRBROTHERS, D. (1961) Serological correspondence between the Cornaceae and Nyssaceae. *Amer. J. Bot.* **48**, 534.

KABAT, E. A. and MAYER, M. M. (1961) *Experimental Immunochemistry*. Chas. C. Thomas, Springfield.

KÄLIN, J. (1946) Die Homologie als Ausdruck ganzheitlicher Baupläne von Typen. *Bull. Soc. Fribourg Sci. Natur.* **37**, 135.

KARR, A. (1849) *Les Quepes*, Janvier, 1849.

KEOSIAN, J. (1964) *The Origin of Life.* Reinhold Pub. Corp., New York.

KERKUT, G. A. (1960) *Implications of Evolution.* Pergamon Press, New York.

KING, T. J. and BRIGGS, B. (1955) Changes in the nuclei of differentiating gastrula cells as demonstrated by nuclear transplantation. *Proc. Nat. Acad. Sci.* **41**, 321.

KIRIAKOFF, S. G. (1959) Phylogenetic systematics versus typology. *Syst. Zool.* **8**, 117.

KIRIAKOFF, S. G. (1962) On the Neo-Adansonian school. *Syst. Zool.* **11**, 180.

KLOZ, J. (1966) Protein characters and their genesis in lower taxons. *Proc. Symposium on the Mutational Process held in Prague,* August 1965, pub. 1966.

KLOZ, J., TURKOVA, V. and KLOZOVA, E. (1963) Legumin and vicilin in some tribes of the family Viciaceae and the systematics of those tribes. *Bull. Serol. Mus.* No. **30**, 1.

KOHN, J. (1957a) A cellulose acetate supporting medium for zone electrophoresis. *Clin. Chem. Acta.* **2**, 297.

KOHN, J. (1957b) An immuno-electrophoretic technique. *Nature* **180**, 986.

KOHN, J. (1960) Cellulose acetate electrophoresis and immuno-diffusion techniques. In Smith, I., editor, *Chromatographic and Electrophoretic Techniques.* Vol. **2**, W. Heinemann, London.

KOHN, J. (1961) Cellulose acetate immuno-diffusion methods. *Bull. Serol. Mus.* No. **26**, 1.

KRAMP, P. L. (1943–1944) On development through alternating generations especially in Coelenterata. *Vidensk. Medd. Naturh. Forening.* Copenhagen, **107**, 13.

KRAUS, R. (1897) Ueber specifische Reaktionen in keimfreien Filtraten aus Cholera, Typhus, und Pestbouillonculturen, erzeugt dorch homologes Serum. *Wien. Klin. Wochenschr.* **10**, 736.

KUBO, K. (1957) Studies on the systematic serology in Sea-Stars I. *J. Faculty of Sci. Hokkaido University Series VI Zool.* Vol. **13**, 67.

KUBO, K. (1958) Studies on the systematic serology of Sea-Stars II. *Annot. Zool. Japon.* **31**, 97.

KUBO, K. (1959a) Studies on the systematic serology of Sea-Stars III. *Annot. Zool. Japon.* **32**, 74.

KUBO, K. (1959b) Studies on the systematic serology of Sea-Stars IV. *Annot. Zool. Japon.* **32**, 214.

KUDO, R. R. (1947) *Pelomyxa carolinensis* Wilson. II Nuclear division and plasmotomy. *J. Morph.* **80**, 93.

KUDO, R. R. (1949) *Pelomyxa carolinensis* Wilson. III Further observations on plasmotomy. *J. Morph.* **85**, 163.

KUDO, R. R. (1960) *Protozoology* (4th edition). C. C. Thomas, Springfield.

KUMMEL, B. (1970) *History of the Earth*, 2nd edition. W. H. Freeman Co., San Francisco.

KWAPINSKI, J. B. (1965) *Methods of Serological Research.* John Wiley & Sons, Inc., New York.

LAKE, G. C., OSBORNE, T. B. and WELLS, H. G. (1914) The immunological relationship of hordein of barley and gliadin of wheat as shown by the complement fixation, passive anaphylaxis and precipitin reactions. *J. Infec. Dis.* **14**, 364.

LAM, H. J. (1936) Phylogenetic symbols, past and present. *Acta Biotheor.* **2**, 153.

LAM, H. J. (1959) Taxonomy—General principles and Angiosperms. In Turrill, W. B., editor, *Vistas in Botany*. Pergamon Press, London.

LAMARCK, J. B. (1809) *Philosophie Zoologique*.

LANDSTEINER, K. (1945) *The Specificity of Serological Reactions*. 2nd revised edition. Harvard University Press, Cambridge.

LANDSTEINER, K. (1962) *The Specificity of Serological Reactions*. Dover Publications, Inc., New York.

LANKESTER, E. R. (1870) On the use of the term homology in modern zoology and the distinction between homogenetic and homoplastic agreements. *Ann. and Mag. Natur. Hist.* **6**, 35.

LANKESTER, E. R. (1899) Zoology in *Encycl. Brit.* **9th** edition. The Werner Co., New York.

LASKEY, R. A. and GURDON, J. B. (1970) Genetic content of adult somatic cells tested by nuclear transplantation from cultured cells. *Nature* **228**, 1332.

LE GROS CLARK, W. E. and MEDAWAR, P. B., editors (1945) *Essays on Growth and Form presented to D'Arcy Wentworth Thompson*. Oxford University Press.

LEONE, C. A. (1949) Comparative serology of some Brachyuran Crustacea and studies in hemocyanin correspondence. *Biol. Bull.* **97**, 273.

LEONE, C. A. (1950a) Serological relationships among common Brachyuran Crustacea of Europe. *Pub. Staz. Zool. di Napoli* **22**, 273.

LEONE, C. A. (1950b) Serological systematics of some Palinuran and Astacuran Crustacea. *Biol. Bull.* **98**, 122.

LEONE, C. (1951) A serological analysis of the systematic relationship of the Brachyuran crab *Geryon quinquedens*. *Biol. Bull.* **100**, 44.

LEONE, C. A. (1952a) Comparative serology and the young systematist. *Bull. Serol. Mus.* No. **8**, 5.

LEONE, C. A. (1952b) Effect of multiple injections of antigen upon the specificity of antisera. *J. Immunol.* **69**, 285.

LEONE, C. (1954) Further serological data on the relationships of some Decapod Crustacea. *Evolution* **8**, 192.

LEONE, C. A. (1968) The immunotaxonomic literature: the Animal Kingdom. *Bull. Serol. Mus.* No. **39**, 1.

LEONE, C. and WIENS, A. (1956) Comparative serology of Carnivores. *J. Mammal.* **37**, 11.

LESTER, R. N., ALSTON, R. E. and TURNER, B. L. (1965) Serological studies in *Baptisia* and certain other genera of the Leguminosae. *Amer. J. Bot.* **52**, 165.

LIBBY, R. L. (1938) The photronreflectometer—an instrument for the measurement of turbid systems. *J. Immunol.* **34**, 71.

LINDEGREN, C. C. (1957a) The role of the gene in evolution. *Ann. N. Y. Acad. Sci.* **69**, 338.

LINDEGREN, C. C. (1957b) Cytoplasmic inheritance. *Ann. N. Y. Acad. Sci.* **68**, 366.

LINOSSIER, G. and LEMOINE, G. H. (1902) Sur la specificité des serums precipitants. *Compt. Rend. de la Soc. de Biol.* **54**, 276.

LORCH, J. (1960) The natural system in biology. St. Catherine's Press, Belgium.

LWOFF, A. (1950) Problems of Morphogenesis in Ciliates. Wiley, N.Y.

MACLEAY, W. S. (1825) Remarks on the identity of certain general laws which have been lately observed to regulate the natural distribution of Insects and Fungi. *Trans. Linn. Soc. Lond.* **14**, 146.

MAGNUS, W. (1908) Weitere Ergebnisse der Serum Diagnostik für die theoretische u. angewandte Botanik. *Ber. d. deutsche botan. Gesellsch.* **26a,** 532.

MANSKI, W., HALBERT, S. P. and AUERBACK, T. P. (1964) Immunochemical analysis of the phylogeny of lens proteins. In Leone, C. A., editor, *Taxonomic Biochemistry and Serology.* Ronald Press Co., New York.

MANTEUFFEL, P. and BEGER, H. (1922) Untersuchungen über unspecifische Reaktionen bei präzipitinierenden Antiseren. *Zeit. f. Immunitatsf. u. exper. Ther.* Teil I Orig. **33,** 348.

MARKERT, C. D. and OWEN, R. D. (1954) Immunogenetic studies of tyrosine specificity. *Genetics* **39,** 818.

MARRACK, J. R. (1938) *The Chemistry of Antigens and Antibodies.* Medical Research Council Special Report Series No. **230.** H.M. Stationery Office, London.

MARTIN, G. W. (1957) Concerning the "cellularity" or "acellularity" of the Protozoa. *Science* **125,** 989.

MASON, H. L. (1957) The concept of the flower and the theory of homology. *Madrono* **14,** 81.

MATOUSEK, J., editor (1965) Blood Groups of Animals. *Proc. 9th European Animal Blood Group Conference.* W. Junk, The Hague.

MCCARTHY, B. J. and BOLTON, E. T. (1963) An approach to the measurement of genetic relatedness among organisms. *Proc. Nat. Acad. Sci.* **50,** 156.

MAYR, E. (1942) *Systematics and the Origin of Species.* Columbia University Press.

MAYR, E. (1965) Numerical phenetics and taxonomic theory. *Syst. Zool.* **14,** 73.

MAYR, E. (1969) *Principles of Systematic Zoology.* McGraw-Hill, New York.

MCELROY, W. D. and GLASS, B., editors (1957) *The Chemical Basis of Heredity.* Johns Hopkins Press, Baltimore.

MCNAIR, J. B. (1935) Angiosperm phylogeny on a chemical basis. *Bull. Torrey Bot. Club* **62,** 515.

MEYER, A. W. (1936) *An Analysis of the De Generatione Animalium of William Harvey.* Stanford University Press.

MEYER-ABICH, A. (1964) The Historico-Philosophical Background of the Modern Evolution-Biology. *Acta. Biotheor.* **2nd supp.** 13.

MICHAELIS, P. (1954) Cytoplasmic inheritance in *Epilobium* and its theoretical significance. *Adv. in Genetics* **6,** 287.

MILLER, S. L. (1953) A production of amino acids under possible primitive earth conditions. *Science* **117,** 258.

MILLER, S. L. (1955) Production of some organic compounds under possible primitive earth conditions. *J. Am. Chem. Soc.* **77,** 2351.

MILLER, S. L. (1957a) The mechanism of synthesis of amino acids by electric discharges. *Biochim. et Biophys. Acta.* **23,** 480.

MILLER, S. L. (1957b) The formation of organic compounds on the primitive earth. In Nigrelli, R. F., editor, *Modern Ideas on Spontaneous Generation. Ann. N.Y. Acad. Sci.* **69,** 260.

MINCHIN, E. A. (1916) The evolution of the cell. *Reports BAAS for 1915,* 437.

MOHAGHEGHPOUR, N. and LEONE, C. A. (1969) An immunologic study of the relationships of non-human Primates to man. *Comp. Biochem. Physiol.* **31,** 437.

MOMENT, G. B. (1945) The relationship between serial and special homology and organic similarities. *Amer. Natur.* **79,** 445.

MONTGOMERY, T. H., JR. (1906) *The Analysis of Racial Descent in Animals.* Henry Holt, New York.

MOODY, P. A. (1958) Serological evidence on the relationships of the musk ox. *J. Mammal.* **39**, 554.

MOODY, P. A., COCHRAN, V. A. and DRUGG, H. (1949) Serological evidence on Lagomorph relationships. *Evolution* **3**, 25.

MOODY, P. A. and DONIGER, D. E. (1965) Serological light on porcupine relationships. *Evolution* **10**, 47.

MOORE, C. R. (1944) Gonad hormones and sex differentiation. *Amer. Natur.* **78**, 97.

MORITZ, O. (1960) Some variants of serological technique developed in serobotanical work. *Bull. Serol. Mus.* No. **24**, 1.

MORITZ, O. (1961) An attempt at a formular description of procedures and results in comparative serology. *Bull. Serol. Mus.* No. **25**, 1.

MORITZ, O. (1964) Some special features of serobotanical work. In Leone, C. A., editor, *Taxonomic Biochemistry and Serology.* Ronald Press Co., N.Y.

MORITZ, O. (1966) Revealing systematical distribution of protein characters by serological methods. In Landa, Z., editor, *Symposium on the Mutational Process.* Symp. CSAV, Praha.

MOSSMAN, H. W. (1937) *Comparative morphogenesis of the fetal membranes and accessory uterine structures.* Carnegie Institution, Washington, Publication **479**, 129.

MULLER, H. J. (1955) Life. *Science* **121**, 1.

MULLER, H. J. (1967) The gene material as the initiator and the organizing basis of life. In Brink, R. A., editor, *Heritage from Mendel.* The University of Wisconsin Press, Madison.

MUNOZ, J. (1954) The use and limitations of serum-agar techniques in studies of proteins. In Cole, W. H., editor, *Serological approaches to Studies of Protein Structure and Metabolism.* Rutgers University Press, New Brunswick, N.J.

NAEF, A. (1926) Zur Diskussion des Homologiebegriffes und seiner Anwendung in der Morphologie. *Biol. Zentralbl.* **46**, 405.

NAEF, A. (1927) Die Definition des Homologiebegriffes. *Biol. Zentralbl.* **47**, 187.

NEUZIL, E. and MASSEYEFF, R. (1958) Parenté immunochimique entre le serum humain et celui de divers animaux; étude immunoelectrophoretique. *Compt. rendus Soc. de Biol.* **150**, 11: 599.

NIGRELLI, R. F., editor (1957) Modern ideas on spontaneous generation. *Ann. New York Acad. Sci.* **69**, 255.

NOBLE, G. K. (1931) *The Biology of the Amphibia.* McGraw-Hill, New York.

NORDENSKIOLD, E. (1928) *A History of Biology.* Tudor Press, New York.

NOWIKOFF, M. (1935) Homomorphie, Homologie und Analogie. *Anat. Anz.* **80**, 388.

NOVIKOFF, M. (1936) L'homomorphie comme base methodologique d'une morphologie comparée. *Bull. Assoc. Russe. Rech. Scientif.*, Prague 4, Sect. sci. nat. et math. (**19**).

NURSALL, J. R. (1962) The origin of the major groups of animals. *Evolution* **16**, 118.

NUTTALL, G. H. F. and DINKELSPIEL, E. M. (1901) On the formation of specific antibodies in the blood following upon treatment with the sera of different animals, together with their use in legal medicine. *J. Hygiene* **1**, 367.

NUTTALL, G. H. F. (1901a) A further note upon the biological test for blood and its importance in zoological classification. *Brit. Med. J.* **2**, 669.
NUTTALL, G. H. F. (1901b) The new biological test for blood in relation to zoological classification. *Proc. Royal Soc. London* **69**, 150.
NUTTALL, G. H. F. (1902) Progress report upon the biological test for blood as applied to over 500 bloods from various sources, together with a preliminary report upon a method of measuring the degree of reaction. *Brit. Med. J.* **1**, 825.
NUTTALL, G. H. F. (1904) *Blood Immunity and Blood Relationship.* Cambridge University Press.
OPARIN, A. I. (1957) *The Origin of Life on the Earth.* (3rd Ed.) Oliver and Boyd, London.
OPARIN, A. I., PASYNASKII, A. G., BRAUNSHTEIN, A. E. and PAULOUSKAYA, T. E., editors (1959), *The Origin of Life on Earth.* Vol. I. Int. Union of Biochem. Symposium Series. Pergamon Press, Great Britain.
O'ROURKE (1959) Serological relationships in the genus *Gadus. Nature* **183**, 1192.
OUCHTERLONY, O. (1949) Antigen-antibody reactions in gels. *Arch. Kem. Mineral Geol. B* **26**, 1.
OUCHTERLONY, O. (1958) Diffusion-in-gel methods for immunological analysis. *Prog. Allergy* **5**, 1.
OUCHTERLONY, O. (1961) Interpretation of comparative immune precipitation patterns obtained by diffusion-in-gel techniques. In Heidelberger, M. and Plescia, O. J., editors, *Immunochemical Approaches to Problems in Microbiology.* Institute of Microbiology, Rutgers University, New Brunswick, N.J.
OUDIN, J. (1946) Methode d'analyse immunochimique par precipitation specifique en milieu gelifie. *Compt. rend. Acad. d. Sci.* **222**, 115.
OUDIN, J. (1952) Specific precipitation in gels and its application to immunochemical analysis. In *Methods in Med. Research* **5**, 335. Yearbook Publ., Chicago.
OWEN, R. (1843) *Lectures on the Comparative Anatomy and Physiology of the Invertebrate Animals.* Longman, Brown, Green, Longmans, London.
OWEN, R. (1847) Report on the archetype and homologies of the Vertebrate skeleton. *Rep. Brit. Assn. Meeting* 1846, **169**.
OWEN, R. (1848) *On the Archetype and Homologies of the Vertebrate Skeleton.* (A reprint of the Brit. Assn. report, revised and extended.) John van Voorst, London.
OWEN, R. (1866) *On the Anatomy of Vertebrates.* Vol. I. Longmans, Green, London.
PANTIN, C. F. A. (1951) Organic design. *Advancement of Science* **8**, 138.
PANTIN, C. F. A. (1960) Diploblastic Animals. *Proc. Linn. Soc. Lond.* **171**, 1.
PANTIN, C. F. A. (1966) Homology, Analogy and Chemical Identity in the Cnidaria. In Rees, W. J., editor, *The Cnidaria and their Evolution.* Symposia of the Zoological Society of London, No. **16**, Published by Academic Press, London and New York.
PAULY, L. (1962) Systematic serology among the Felidae and other closely related groups. *Bull. Serol. Mus.* No. **28**, 5.
PAULY, L. and WOLFE, H. R. (1957) Serological relationships among members of the order Carnivora. *Zoologica* **42**, 159.
PEARSE, A. S. (1947) *Zoological Names. A List of Phyla, Classes, and Orders.* 2nd Edition. Prepared for Section F, AAAS. Durham, North Carolina.

PEARSE, A. S. (1949) *Zoological Names. A List of Phyla, Classes, and Orders.* 4th Edition. Prepared for Sect. F, AAAS.

PICKEN, L. (1960) *The Organization of Cells and Other Organisms.* Oxford at the Clarendon Press.

PITELKA, D. R. (1963) *Electron-microscopic Structure of Protozoa.* Pergamon Press, New York and London.

PONTECORVO, G. (1959) *Trends in Genetic Analysis.* Columbia University Press. Oxford University Press.

RACE, R. R. and SANGER, R. (1968) *Blood Groups in Man.* 5th edition. F. A. Davis Co., Philadelphia.

RANKAMA, K. (1954) *Isotope Geology.* Pergamon Press, London.

REES, W. J. (1957) Evolutionary trends in the classification of capitate hydroids and medusae. *Bull. Brit. Mus. (Nat. Hist.) Zool.* **4**, 453.

REES, W. J. (1966) *The Cnidaria and their Evolution.* Zool. Soc. Lond. Academic Press, London.

REMANE, A. (1955) Morphologie als Homologienforschung. *Zool. Anz.*, Supp. **18**, 159.

REMANE, A. (1956) *Die Grundlagen des natürlichen Systems, der vergliechenden Anatomie und der Phylogenetik.* 2nd edition. Akad. Verlagsges, Leipzig.

REMANE, A. (1963) The evolution of the Metazoa from colonial flagellates vs. plasmodial ciliates. In Dougherty, E. C., editor (1963), *The Lower Metazoa.* University of California Press, Berkeley.

ROJERS, D. J. and TOMINOTO, T. T. (1960) A computer program for classifying plants. *Science* **132**, 1115.

ROTHSCHILD, Lord (1961) *A Classification of Living Animals.* Longmans, Green & Co., New York.

ROWSON, J. M. (1958) Symposium on biochemistry and taxonomy. Alkaloids in plant taxonomy. *Proc. Linn. Soc. Lond.* **169**, 212.

RUSSELL, N. H. (1962) The development of an operational approach in plant taxonomy. *Syst. Zool.* **10**, 159.

SAINT-HILAIRE, G. (1818) *Philosophie Anatomique.* Paris.

SARICH, V. M. and WILSON, A. C. (1966) Quantitative immunochemistry and the evolution of Primate albumins: micro-complement fixation. *Science* **154**, 1563.

SARICH, V. M. and WILSON, A. C. (1967a) Rates of albumin evolution in Primates. *Proc. Nat. Acad. Sci.* **58**, 141.

SARICH, V. M. and WILSON, A. C. (1967b) Immunological time scale for hominid evolution. *Science* **158**, 1200.

SATTLER, R. (1964) Methodological problems in taxonomy. *Syst. Zool.* **13**, 19.

SCHECHTMAN, A. M. and NISHIHARA, T. (1955) The cell nucleus in relation to the problem of cellular differentiation. *Ann. N.Y. Acad. Sci.* **60**, 1079.

SCHEINBERG, S. L. (1960) Genetic studies of serum antigens in species hybrids. *Genetics* **45**, 173.

SCHINDEWOLF, O. H. (1969) Homologie und Taxonomie. In Anniversary Volume dedicated to Professor Dr. C. J. Van der Klaauw. *Acta Biotheoretica* **18**, 235.

SCHMALHAUSEN, I. I. (1949) *Factors of Evolution. The Theory of Stabilizing Selection.* The Blakiston Co., Philadelphia.

SCHMIDT, J. (1922) The breeding places of the eel. *Phil. Trans. Roy. Soc.* B. **211**, 179.

SCHOPF, J. W. (1970) Pre-Cambrian micro-organisms and evolutionary events prior to the origin of vascular plants. *Biol. Rev.* **45**, 319.

SCHRADER, F. (1953) *Mitosis*. (2nd edition.) Columbia University Press, New York.

SHULL, A. F. (1929) *Principles of Animal Biology*. McGraw-Hill, New York.

SHUSTER, C. N. (1962) Serological correspondence among horseshoe "Crabs" (Limulidae). *Zoologica* **47**, 1.

SIBLEY, C. G. (1960) The electrophoretic patterns of avian eggwhite proteins as taxonomic characters. *Ibis* **102**, 215.

SIBLEY, C. G. and JOHNSGARD, P. A. (1959) An electrophoretic study of eggwhite proteins in twenty-three breeds of the domestic fowl. *Amer. Natur.* **93**, 107.

SILVER, L. T. (1970) Uranium, Chorium-lead isotope relations in lunar materials. *Science* **167**, 468.

SIMPSON, G. G. (1945) The principles of classification and a classification of mammals. *Bull. Amer. Mus. Natur. Hist.* **85**, 213.

SIMPSON, G. G. (1949) *The Meaning of Evolution*. Yale University Press.

SIMPSON, G. G. (1953) *The Major Features of Evolution*. Columbia University Press.

SIMPSON, G. G. (1959) Anatomy and morphology: Classification and evolution; 1859 and 1959. *Proc. Amer. Philos. Soc.* **103**, 286.

SIMPSON, G. G. (1961) *Principles of Animal Taxonomy*. Columbia University Press, New York.

SINGER, C. (1959) *A History of Biology*. 3rd edition. Abelard-Schuman, London and New York.

SOKAL, R. R. and SNEATH, P. H. A. (1963) *Principles of Numerical Taxonomy*. W. H. Freeman & Co., San Francisco.

SOKAL, R. R. (1972) Taxonomy rationalized. *Nature* **236**, 412.

SONNEBORN, T. M. (1950) The cytoplasm in heredity. *Heredity* **4**, 11.

SONNEBORN, T. M. (1957) Breeding systems, reproductive methods, and species problems in Protozoa. In Mayr, E., editor, *The Species Problem*. Pub. No. 50, AAAS., Washington, D.C.

SONNEBORN, T. M. (1963a) Bearing of Protozoan studies on current theory of genic and cytoplasmic actions. *Proc. 16th Int. Congr. Zool.* **3**, 197.

SONNEBORN, T. M. (1963b) Does preformed structure play an essential role in cell heredity? In Allen, J. M., editor, *The Nature of Biological Diversity*. McGraw-Hill, New York.

SONNEBORN, T. M. (1967) The evolutionary integration of the genetic material into genetic systems. In Brink, R. A., editor, *Heritage from Mendel*. University of Wisconsin Press, Madison, Wisconsin.

SPEMANN, H. (1915) Zur Geschichte und Kritik des Begriffs der Homologie. *Die Kultur der Gegenwart Teil.* **3** Abt. 4, Bd. 1.

STALLCUP, W. B. (1954) Myology and serology of the avian family Fringillidae, a taxonomic study. *Univ. Kansas Pub. Mus. Natur. Hist.* **8**, 157.

STALLCUP, W. B. (1961) Relationships of some families of the suborder Passeres (songbirds) as indicated by comparisons of tissue proteins. *J. Grad. Res. Center. So. Methodist Univ.* **29**, 43.

STEBBINS, G. L. (1960) The comparative evolution of genetic systems. In Tax, S., editor, *Evolution after Darwin*. University of Chicago Press.

STEENSTRUP, J. J. S. (1845) On the Alternation of Generations or the Propagation and Development of Animals through Alternate Generations. *Ray Soc.* **7**, London.

STRATTON, F. and RENTON, P. H. (1958) *Practical Blood Grouping*. Blackwell, Oxford.

STRICKLAND, H. E. (1844) Report on the recent progress and present state of ornithology. *Rep. Brit. Assn. Adv. Sci.* 14, 170. York Meeting. London (1845).

STRICKLAND, H. E. (1846) On the structural relations of organized beings. *Phil. Mag.* 3rd series 28, 354.

SUOMALAINEN, E. (1950) Parthenogenesis in animals. *Adv. Genet.* 3, 193.

SWAINSON, W. (1835) *The Geography and Classification of Animals*. Longman, et al. and Taylor. London.

TAX, S., editor (1960), *Evolution after Darwin*. University of Chicago Press.

TCHISTOVITCH, Th. (1899) Études sur immunisation contres le serum d'anguilles. *Ann. de l'Inst. Pasteur* 13, 406.

THOMPSON, D'ARCY W. (1951) *On Growth and Form*. Reprinted from 2nd edition. 1942. Cambridge University Press.

THOMPSON, W. R. (1960) Systematics: the ideal and the reality. *Studia Entom.* 3, 493.

THOMPSON, W. R. (1962) Evolution and taxonomy. *Stud. Entomol.* 5, 549.

TOBISKA, J. (1964) *Die Phytohämagglutine*. Akademie-Verlag. Berlin.

TUCKER, D. W. (1959) A new solution to the Atlantic eel problem. *Nature*, 183, 495.

TURNER, W. (1899) Anatomy. In *Encycl. Brit.* 9th edition. Vol. I. The Werner Co., New York.

UHLENHUTH, P. (1901) Weitere Mittheilungen uber meine Methode zum Nachweise von Menschenblut. *Deut. med. Wochensch.* 27, 260.

UHLENHUTH, P. (1903) Zur historischen Entwickelung meines forensischen Verfahrens zum Nachweis von Blut und Fleisch mit Hüle spezifischer Sera. *Deut. tierärztl. Wochenschr.* 11, No. 16.

UREY, H. C. (1952a) *The Planets*. Yale University Press, New Haven.

UREY, H. C. (1952b) On the early chemical history of the earth and the origin of life. *Proc. Nat. Acad. Sci.* 38, 351.

UREY, H. C. (1963) *Some Cosmochemical Problems*. Notes on the Thirty-seventh Annual Priestley Lectures. Edited and sponsored by MU chapter of Phi Lambda Upsilon and associated Departments. Penn. State University, University Park, Pa.

VAN NEIL, C. B. (1956) The classification and natural relationships of bacteria. *Symp. Quant. Biol.* 11, 285. Cold Spring Harbor.

VAVILOV, N. I. (1922) The law of homologous series in variation. *J. Genet.* 12, 47.

VIGORS, N. A. (1825) Observations on the natural affinities that connect the orders and families of birds. *Trans. Linn. Soc. Lond.* 14, 395.

VINOGRADOV, A. P. (1960) The origin of the biosphere. In Florkin, M., editor, *Aspects of the Origin of Life*. Pergamon Press, London.

VORONTSOVA, M. A. and LIOSNER, L. D. (1960) *Asexual Reproduction and Regeneration*. Pergamon Press, London.

WAGNER, R. P. and MITCHELL, H. K. (1955) *Genetics and Metabolism*. Wiley & Sons, New York.

WALLACE, D. G., MAXSON, L. R. and WILSON, A. C. (1971) Albumin evolution in frogs: a test of the evolutionary clock hypothesis. *Proc. Nat. Acad. Sci.* 68, 3427.

WEIGLE, W. O. (1962) Biological properties of the cross-reactions between anti-BSA and heterologous albumins. *J. Immunol.* 88, 9.

WELLS, H. G. (1929) *The Chemical Aspects of Immunity*. Amer. Chem. Soc. Monograph Series, New York.

WEMYSS, C. T., JR. (1953) A preliminary study of Marsupial relationships as indicated by the precipitin test. *Zoologica* **38**, 173.

WERNER, B. (1963) Effect of some environmental factors on differentiation and determination in marine Hydrozoa, with a note on their evolutionary significance. *Ann. N.Y. Acad. Sci.* **105**, 463.

WHITE, M. J. D. (1954) *Animal Cytology and Evolution*, 2nd edition. Cambridge.

WILHELMI, R. W. (1940) Serological reactions and species specificity of some Helminths. *Biol. Bull.* **79**, 64.

WILHELMI, R. W. (1942) The application of the precipitin technique to theories concerning the origin of vertebrates. *Biol. Bull.* **82**, 179.

WILHELMI, R. W. (1944) Serological relationships between the Mollusca and other invertebrates. *Biol. Bull.* **87**, 96.

WILLIAMS, C. A., JR. and GRABAR, P. (1955) Immunoelectrophoretic studies on serum proteins. I. The antigens of human serum. *J. Immunol.* **74**, 158.

WILLIAMS, C. A., JR. (1956) Antigen antibody reactions in gels and systematic serology. *Proc. 14th Int. Zool. Congr.*, Copenhagen, 1953, 336.

WILLIAMS, C. A., JR. (1964) Immunochemical analysis of serum proteins of the Primates. A study in molecular evolution. In Buettner-Janusch, J., editor, *Evolutionary and Genetic Biology of the Primates*, Vol. **II**, Academic Press, New York.

WILLIAMS, C. A., JR. and WEMYSS, C. T., JR. (1961) Experimental and evolutionary significance of similarities among serum protein antigens of man and the lower Primates. *Ann. N.Y. Acad. Sci.* **94**, 77.

WILLIS, J. C. (1922) *Age and Area*. Cambridge University Press.

WILLIS, R. (1847) *The Works of William Harvey, M.D.* Sydenham Society, London.

WILSON, E. B. (1896) The embryological criterion of homology. *Biol. Lect. for 1894*. Marine Biol. Lab., Ginn & Co., Boston.

WILSON, E. B. (1925) 2nd edition. *The Cell in Development and Heredity*. Macmillan, New York.

WITHERS, R. F. J. (1964) Morphological correspondence and the concept of homology. In Gregg, J. R. and Harris, F. T. C., editors, *Form and Strategy in Science*. D. Reidel Publishing Co., Holland.

WOLFE, H. R. (1935) The effect of injection methods on the species specificity of serum precipitins. *J. Immunol.* **29**, 1.

WOLLMAN, E. L., JACOB, F. and HAYES, W. (1956) Conjugation and recombination in *Escherichia coli* K–12. *Cold Spring Harbor Symp. Quant. Biol.* **21**, 141.

WOODGER, J. H. (1929) *Biological Principles—A Critical Study*. Routledge and Kegan Paul, London.

WOODGER, J. H. (1945) On biological transformations. In *Essays on Growth and Form*, edited by W. E. Le Gros Clark and P. B. Medawar, Oxford University Press.

WOODGER, J. H. (1952) *Biology and Language*. Cambridge University Press.

WRIGHT, S. (1934) Genetics of abnormal growth in the Guinea Pig. *Cold Spring Harbor Symp. Quant. Biol.* **2**, 137.

WRINCH, D. (1941) The native protein theory of the structure of cytoplasm. *Cold Spring Harbor Symp. Quant. Biol.* **9**, 218.

ZANGERL, R. (1948) The methods of comparative anatomy and its contribution to the studies of evolution. *Evolution* **2**, 351.

ZEUNER, F. (1958) *Dating the Past.* Methuen, London.

ZIMMERMAN, W. (1931) Arbeitsweise der botanischen phylogenetik und anderen Gruppierungswissenschaften. In Abderhalden, *Handbuch der biologischen Arbeits Methoden.* Abt. **3**, 2, Teil 9, 941.

ZIMMERMAN, W. (1954) in Heberer, G., editor, *Die Evolution der Organismen. Methoden du Phylogenetik.* 2nd edition. Jena.

AUTHOR INDEX

Akabori, S. 22
Allfrey, V. G. 100
Alston, R. E. 190
Ascoli, M. 146

Bacon, F. 214
Baer, K. E. von 119
Baier, H. L. 148
Baier, J. G. 175
Baker, J. R. 8, 9, 10
Barghoorn, E. S. 24, 37
Basford, N. 175, 196, 197
Bassindale, R. 215
Bateson, W. 127
Bather, F. A. 231
Baum, W. 175
Beadle, G. W. 74, 85
Beger, H. 184
Belon, P. 77
Berkner, L. V. 19, 20, 40, 58
Blackler, A. W. 100, 101
Blackwelder, R. E. 108, 205, 207,
    227, 229, 244, 250, 255
Blyth, E. 74
Bock, W. J. 12, 83, 118, 137, 139
Bolton, E. T. 101, 151, 155, 192, 198–
    199, 237–239, 254
Bonner, J. T. 67, 94, 95, 98
Boor, A. K. 185
Bordet, J. 144, 179
Bowley, C. C. 143
Boyd, W. C. 143, 150, 185, 186
Brachet, J. 100
Breder, C. M. Jr. 26
Breuer, M. E. 88
Briggs, R. 85, 86
Brink, R. A. 87, 96, 97
Britten, R. J. 90, 239
Buchsbaum, R. 60, 61, 62
Butler, J. E. 175, 189, 193

Cain, A. J. 209, 210, 223, 229, 242,
    255
Camin, J. H. 254
Camper, P. 77
Cawley, L. P. 158
Cei, J. M. 175
Clark, A. H. 35
Cleveland, L. R. 34
Cochran, V. 176
Cohen, E. 175
Cole, W. H. 156, 157
Commoner, B. 23
Coonradt, V. 74
Corliss, J. O. 8, 9, 10
Cracraft, J. 140
Crew, F. A. E. 105
Crosby, J. L. 29
Crowle, A. J. 158
Crowson, R. A. 205, 227, 230, 253
Cuvier, G. 81, 119, 122

Daily, B. 24
Danielli, J. F. 93, 94, 98, 99
Darlington, C. D. 1, 2, 34, 69
Darlington, P. J. Jr. 250, 251, 254
Darwin, C. 1, 49, 81–84, 106, 110,
    118, 122, 126, 127, 137, 209–220,
    222, 223, 225, 235, 243, 244, 247,
    253, 254, 256
Darwin, F. 211, 216
Dawes, B. 1
De Beer, G. R. 35, 42, 65, 66, 77, 82,
    127, 133, 240
De Falco, R. J. 146, 175
Delevoryas, T. 230, 253
Dessauer, H. 201
De Vries, H. 239
Dobell, C. C. 5, 6, 8, 9, 10, 30, 110
Dobzhansky, Th. 30, 245
Doniger, D. 176

277

SUBJECT INDEX